REACHING FOR THE STARS
INDIA'S JOURNEY TO MARS AND BEYOND

To Deborah,
With Best Wishes
to the best friend
of science journalists.

Pallava Bagla
June 9, 2015

Modern science meets ancient technology. This unique flying saucer shaped building at India's spaceport Sriharikota houses the sophisticated Mission Control Centre. Seen in the foreground is the country's iconic Ambassador car.

Credit: Pallava Bagla

REACHING FOR THE STARS
India's Journey to Mars and Beyond

PALLAVA BAGLA
SUBHADRA MENON

BLOOMSBURY
LONDON • NEW DELHI • NEW YORK • SYDNEY

BLOOMSBURY PUBLISHING INDIA PVT. LTD.
London   New Delhi   New York   Sydney

ISBN: 978-93-84052-32-4

10 9 8 7 6 5 4 3 2 1

Published by Bloomsbury Publishing India Pvt. Ltd.
Vishrut Building, DDA Complex, Building No. 3
Pocket C-6 & 7, Vasant Kunj
New Delhi 110 070

Printed at ANVI COMPOSERS, Paschim Vihar, New Delhi

Front end paper (double spread): Universal splendour: The magical inter-planetary alignments of the solar system. *Credit:* NASA

Facing half-title page: Dance of the planets, an artist's impression. *Credit:* NASA

Map on page 259: The dramatic expanse of India, from Kashmir to Kanya Kumari and the Rann of Kutch to the Sunderbans, as seen from space. This cloud-free composite image of the subcontinent was captured by the country's indigenously made Resourcesat 1. *Credit: ISRO*

Back end paper (double spread): A mural from Darya Daulat Sultan Bagh in Srirangapatna near Bangalore, celebrating the victory of Hyder Ali and Tipu in 1780 over the British. Col. Baillie is seen seated in a palanquin in the middle. An ammunition tumrel on the top left corner of the square is seen to have caught fire by a rocket attack according to several reports. *Credit: Pallava Bagla*

Dedication

To Nayantara and Ashwat, our wonderful children, without whom this book, and much of our respective professional lives, would not have been a reality. Their unconditional support and love have made all the difference. Words are inadequate to express our gratitude to them, and to their own singular intelligence and art of questioning that keeps us all on a constant quest for new knowledge and discovery. Go, both of you, reach for the stars in life.

Night life over India. This image, taken by satellites flown by NASA shows the night lights over the Indian subcontinent. Looking at this image can anybody imagine that 400 million Indians still do not have access to electricity?

Credit: NASA

Contents

ISRO trained its most sophisticated Earth viewing satellite Cartosat-2 to take this spectacular view of the Moon.

Credit: ISRO

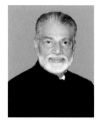

Dr K Radhakrishnan
Chairman
Indian Space Research Organization

Foreword

The exploration of space is a fascinating and exciting field of science and technology. India is a relatively senior and well-respected member of a group of nations that possesses capability to build space systems, launch space missions from its own soil and pass benefits on to the people. In continuation of this eventful and momentous journey, a significant milestone was recently crossed, and that is the Mars Orbiter Mission (MOM) as India's maiden inter-planetary sojourn, conceptualised, designed and executed by India. Moreover, it is a mission that was put together in record time, taking less than one and a half years from start to launch. This has been possible because of the strong intent and support from the government, extraordinary intrinsic diligence and rapid decision-making processes within the Indian Space Research Organization (ISRO), as well as commitment from a wide range of people from its many partner institutions.

It goes without saying that the Mars Orbiter Mission is an exciting story for many, because it is about how ISRO took on this complex and challenging task of designing a long inter-planetary journey for a small satellite using its trusted Polar Satellite Launch Vehicle and carrying with it payloads that can help India contribute to the global pool of knowledge and information about the Earth's next door planetary neighbour, Mars.

The Mission has been, in some ways, a pinnacle of achievement for ISRO, although the organisation is known for not resting on its laurels and moving rapidly into the future with fresh plans for new missions, explorations and applications. With this mission, we have proved our capabilities in multiple areas of space technology: launch vehicles for long, inter-planetary flights, scientific payloads designed within stringent weight and cost restrictions, a spacecraft that can finally settle into the Martian orbit and function with complete on-board autonomy given the challenges of communicating over great distances, and an extensive ground network that can help navigate, track and capture data for this important mission.

This book captures all of this and more, and fills a gap in that sense. It tells the story of MOM, and of ISRO's experiences in designing and executing this mission with complete indigenous capability. It dwells on the international scene and ISRO's future plans, where we hope to take our Moon mission's success to the next level with Chandrayaan 2, a mission to study the Sun, Aditya, and develop essential technologies to send Indian crew to outer space. The Geosynchronous Satellite Launch Vehicle (GSLV) Mark III is on the way to the launch pad for an experimental mission, towards a significant milestone that will change the way we set limits to our current launch vehicle capabilities.

Thousands of people from across the country have been involved in this Mars Orbiter Mission, a Team ISRO effort in every sense of the word. In fact, it is the kind of team work we coordinate that sets ISRO apart, besides the fact that every target is clearly set and there is little ambiguity in what we want to achieve in a set time period.

We have often been challenged in trying to satisfy our critics who make suggestions that space technologies are unnecessary luxuries in a country like ours where people go hungry each day even today and that too much is being spent on the space programme. To this I can only say that we have to view all this in a national perspective. What India is spending for its space programme is just 0.34 per cent of its Central Government expenditure and of that nearly 7–8 per cent is spent on space exploration. Meanwhile, the benefits are tremendous, even if you just look at the impact it has made on society through disaster management, since till recently it was common for 10,000 lives to be lost in a year because of cyclones and other such disasters, but now, with new and useful satellite technology and new applications for people and society, there is a huge difference. Using space applications we can follow the movement of a cyclone and warn people through our own system. Take the agriculture scenario, where space-based forecasts well in advance of harvesting are helping the government take informed decisions. These are but a few things. You look at the communications scenario – in the 1980s television was available in hardly four metro cities in the country, and there were no cell phones. These are revolutions, nothing less. Today, the Mars Orbiter Mission has brought India to a new level: demonstrating a novel approach to developing the capability for inter-planetary missions, within the envelope of current capabilities.

It was therefore an honour to share the different dimensions of this story with one of India's most respected and senior science journalists, Pallava Bagla, whom I have known and interacted with for several years now. As a practicing engineer and the Chairman of ISRO, I can say without doubt that sharing complex details about the mission with Mr Bagla over the time of its development and launch and in the post-launch period was a pleasure simply because one is confident that they are not just being appreciated but also understood, and that they would be faithfully retold to larger audiences as part of the task of public engagement on such a critical issue. Dr Subhadra Menon, also an award-winning author with a deep understanding of science and technology institutions and the scope and practice of science in India, spent a great deal of time visiting our facilities, interacting with our leadership and engineers who have worked tirelessly on this mission.

Certainly the Mars mission is going to create a great sense of pride and achievement and it is also going to improve our position in the international community.

September 2014

K Radhakrishnan
Chairman, ISRO

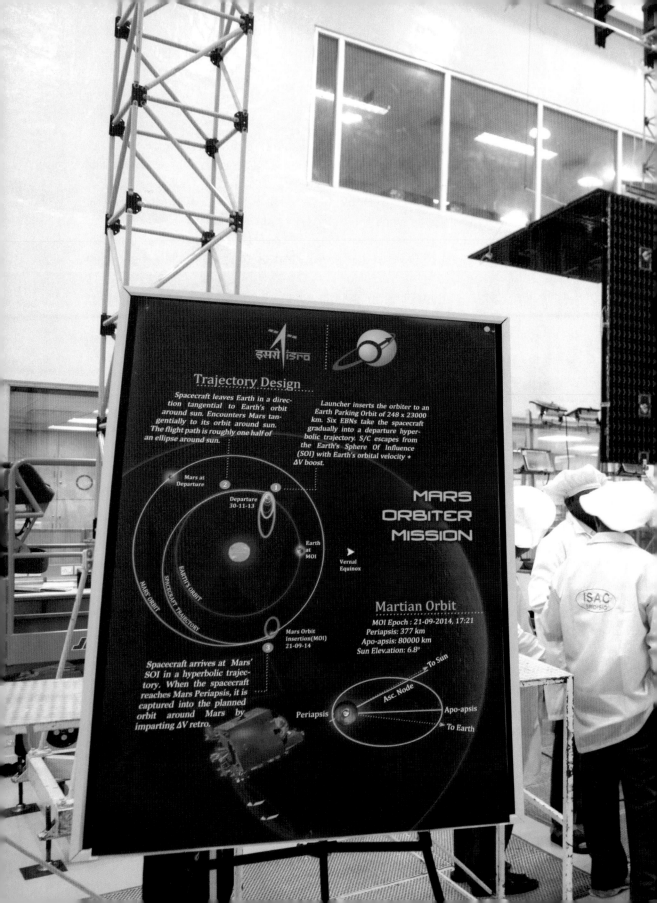

The Mars Orbiter nearing completion: This was a 15-month roller coaster ride for the satellite makers at ISRO.

Credit: Pallava Bagla

The red planet, Mars, of allure and mystery.
Credit: NASA

Preface

You are what your deep, driving desire is. As your desire is, so is your will. As your will is, so is your deed. As your deed is, so is your destiny.
                                                          – Brihadaranyaka Upanishad

Reaching for the stars is literally every child's dream.

In every nation's firmament of history, some stars shine brighter than the rest. These are the outliers – the institutions, collectives or individuals who seem to constantly push their limits of excellence, efficiency and productivity. They defy the inertness that seems to grip others and create a magical productivity out of the same rules and guidelines of the democratic polity that governs a country. Their milestones and achievements seem to fly in the face of the inadequacies that otherwise appear to cripple the public system in many realms of development. This truism is very prominent in a country like India, and the Indian space programme is one of these outliers. 'This is one domain in which we are at the international cutting edge. A domain in which we have pushed beyond mediocrity to achieve excellence,' said Prime Minister Narendra Modi while witnessing a copy-book launch at the Indian spaceport of Sriharikota in Andhra Pradesh in June, 2014. This excellence has, naturally, led to extraordinary things. Among these is the first, deep space inter-planetary mission to Mars, conceptualised, planned and implemented by the Indian Space Research Organization (ISRO), a mission made remarkable not just by the very idea, but by the fact that ISRO ran it like a sprint, getting everything ready within a record 15 months, on a shoe string budget of ₹4,500 million (USD 70 million), which, PM Modi reminded the country, is less than what Hollywood spent on the sci-fi thriller *Gravity*.

So it seemed like there was a story to tell, of the dramatic realisation of a dream when a smart little, Tata Nano car-sized satellite weighing 1,350 kilograms hurtled out into deep space after starting off in November 2013, on a journey of 690 million kilometres and 300 days. A trial of ISRO's technological ability for a long journey beyond the Earth's gravitational field for the first time, the Mars Orbiter Mission (MOM) also attempts to probe further into whether we are indeed alone in the Universe or not. The geopolitics is significant too, with the timing helping India race ahead of China in trying to get to Earth's alluring neighbour. National pride is always a driver for such missions. Perhaps the world's cheapest inter-planetary mission, MOM was a tense roller-coaster for the teams involved, some deadlines so stiff that the mission almost never took off, but for the never-say-die spirit of the teams at ISRO. It became, therefore, an eye-opener for the world.

Mangalyaan was first publicly announced by former Prime Minister Dr Manmohan Singh from the magnificent Red Fort in old Delhi on August 15, 2012. In figuring out the orbital callisthenics special to deep space journeys, the incredibly long-distance communication required, and of course fashioning the little boxy payloads that would try to answer some of the mission's scientific objectives, the well-known Indian ability to innovate with frugal engineering or *jugaad* can be observed! More than 500 engineers and scientists, from the satellite makers at ISRO's headquarters in Bangalore, to the scientists at the Space Applications Centre (SAC) and the Physical Research Laboratory (PRL) in Ahmedabad, to the rocket fabricators quietly beavering away in the emerald greenery of the Vikram Sarabhai Space Centre (VSSC) in Thiruvananthapuram, and those manning the ground stations, are directly a part of the mission.

This book is a narrative by two authors, ideated through travels and close encounters with those who run the space programme from the vast and spreading network of ISRO institutions. When you take in the air at VSSC, very much the cradle of the Indian space programme, the early energy and passion of Sarabhai and others washes over you. Walking through the corridors of what are otherwise indistinctive, government-style buildings, it is not difficult to understand that this positive energy and the vision set up decades ago is still alive, even the reason why success has become a habit here. This book is an insider's narrative, in every sense of the word, capped with a deep and personalised view of the Indian space programme resulting from years of intensive and impassioned news reporting. Through this time, what has touched the heart and set off a spark of excitement is how ISRO and the space programme seem to straddle high-tech India and the grass-roots Bharat. Here are space engineers and scientists who dabble each day with an internationally competing space programme while remaining rooted in age-old traditions and coming from simple, small towns of India. ISRO's islands of excellence live in a vast ocean of Indians eking out a living from the land. Both benefit from each other in a way, depending on each other, and life goes on. But it is really the deep contrast between the two that makes telling of the story not just exciting and all-consuming, but philosophically challenging too. '*MOM views Mother India!*' was the first caption used to detail some of the early pictures that were sent back by the Mars Orbiter as it began its journey. This somehow evoked a lot of national pride, and having lamented earlier on why

ISRO had decided against having the Indian flag on the satellite, one felt even greater regret about its absence from the craft.

Within just half a century, India has developed what is called end-to-end capability – from building its own heavy-duty rockets to fabricating sophisticated satellites and payloads, much of it at a point in time when technology denials were the norm. The Mars Orbiter satellite is a complex piece of electronics on its own, carrying five scientific payloads, each one completely indigenous. On its 24th consecutively successful mission, the Polar Satellite Launch Vehicle rocket is about the same weight (320 tons) as a fully

Up close and personal: The author, Pallava Bagla, seen here standing next to the Mars Orbiter satellite ready for its long journey to the Red Planet. Pallava's fingerprints are literally on the Mangalyaan satellite! A rare privilege indeed.

Courtesy: Pallava Bagla

loaded Boeing Jumbo Jet or 50 fully-grown elephants, as tall as a 15-storey building. From the Thumba church rocket lab where India's former and famous President APJ Abdul Kalam made his first forays, to the Goliath that rolls out to Sriharikota from VSSC in contemporary India, these have been long journeys. Kalam, in a sense, is the bridge between the two schools of rocketry in India, one being ISRO, and the other the Defence Research Development Organisation that makes missiles.

How often is it that people try to overcome their dejection or angst by just sitting and staring at night skies – as though they seek hope and help from the black nothingness of the Universe. That is why space is perceived as both magical and comforting. Then, catalysed by the charismatic, chiselled diligence of the trailblazer Vikram Sarabhai and many others who led from the front – Satish Dhawan, UR Rao, K Kasturirangan, G Madhavan Nair, and K Radhakrishnan – the Indian space programme has increasingly been more about societal application actively contributing to the necessities and comforts of contemporary life. 'No ambiguity of purpose', as Sarabhai famously remarked in defence of why a country like India needed to get into space exploration. The critics are still around, and so are the dreamers, but most importantly, the doers are there too. Inter-planetary missions like the Mars mission, a second Moon mission, Chandrayaan-2, that hopes to land a rover on the Moon's surface, a flight to the Sun using Aditya and the ambition of putting Indians out in space through the Human Spaceflight Programme, India's space programme has an action-packed future. Hopefully, so does India. We may or may not be around to chronicle many of the forthcoming successes or failures, but hopefully this book will leave behind some lasting narratives of the romance and drama that India experienced on its journey to the Moon, Mars and beyond.

MOM has a competitor in reaching Mars. NASA's MAVEN spacecraft (Mars Atmosphere and Volatile Evolution) launched on November 18, 2013 is a USD 672 million mission to study the atmosphere of Mars. Seen here is the launch of the Atlas V rocket at the Cape Canaveral launch pad in Florida, USA.

Credit: NASA

Acknowledgements

A book like this becomes a reality only because countless people give generously of their time, sharing knowledge and diverse experiences without a moment's hesitation, and providing facts, anecdotes and details with a certain selflessness that strengthens your faith in human goodness. Books also become a reality because some very special people are always around, with encouragement, love, buffer and unflagging trust.

First of all, we are deeply appreciative of several officials at ISRO. K Radhakrishnan, who has spent so much of his time explaining complexities and sharing context despite what must be a gruelling schedule for the leader of the pack. Kiran Kumar, MYS Prasad, VK Dadhwal, JN Goswami, SK Shivakumar, TK Alex, M Chandradathan, M Annadurai, S Arunan, P Kunhikrishnan, PG Diwakar, BN Suresh, Deviprasad Karnik, Somya Sarkar, Vijayamohan Kumar, VS Hegde, Shantanu Bhatavadekar, Seshagiri Rao, Nilanjan Raut, Anil Bharadwaj, Ashutosh Arya, S Somanath, S Unnikrishnan Nair and Yashodha RY. Our deep appreciation to Guruprasad, whose encyclopaedic knowledge about the Indian and global space programmes is incredible, and to Mukund Rao, Sridhar Murthy, VP Balagangadharan, S Satish. Former Chairmen of ISRO, G Madhavan Nair, K Kasturirangan and UR Rao deserve our special appreciation.

To astronaut Rakesh Sharma, our sincere thanks for his unique insights into human space flight. Thanks are due to SK Malhotra for support at all times, and to Ravi Gupta, Pankaj Pachauri, Syed Akbaruddin, Sharat Chander and Neelam Kapur.

To General Charles F Bolden we express our thanks, as we do to the public relations staff at NASA, David S Weaver, Allard A Beutel, Michael Breukus, Dennis McSweeney and Guy W. Webster. From the European Space Agency, we thank Nadia Imbert-Vier.

We express our very special gratitude to Surit Mitra, the consummate 'books' man, without whose spontaneous and wise support we may have been in a very different zone as far as publishing goes.

We consider ourselves very fortunate to have received advice, guidance and goodwill of several great professionals from diverse fields, who through their work and gracious advice and sharing, have had a major influence on our work. We thank CNR Rao, President APJ Abdul Kalam, R Chidambaram, Anil Kakodkar, Raghunath Mashelkar, Ravi B Grover, K Srinath Reddy, Raj Chengappa, Felicity Young, Tony Bondurant, Srikumar Banerjee, SK Saraswat and Avinash Chander.

We are very grateful to National Security Advisor Ajit Doval, Ambassador Shivshankar Menon and Minister Jitendra Singh. Also to K Vijayaraghavan, Shailesh Nayak, Kiran Karnik and Aiyagiri Rao, we record our special appreciation. To VS Ramamurthy and Baldev Raj, a warm thanks for giving the strength. To Mallika Sarabhai, with gratitude always.

We are very appreciative of the support from New Delhi Television (NDTV), especially colleagues Sonia Singh, Barkha Dutt, Chetan Bhattacharji, Manika Raikwar, Suparna Singh and Sunil Saini, all incredibly alert gate keepers, but with a human touch. Also to Vikram Chandra, Ravish Kumar, Amitabh Revi, Deepak Kumar Choubey, Nitin Gokhale, Maya Sharma, Ajmal Jami, Anamitra Chakladhar, Sam Daniel, Radhika Iyer, Nilanjana Bhaduri-Jha, Mayanka Kapoor, Sudhakar, Madhusudan Srinivas, Pari S Vallal, Pranav Duttgautam, Alphonse Raj, N. Sukumar, Alexander Edwin, K Suresh and Devraj without whom a lot of the space stories would not have received the attention they deserved. It is a delight to work with such a talented team while gathering all the news that is fit to broadcast. Vishnu Som, for his singular command over all man-made things that fly or swim, and his simply superb anchoring on television. And finally, at NDTV, Prannoy Roy and Radhika Roy – you are the best!

At *Science* magazine, warm thanks to Marcia McNutt, Tim Appenzeller, Richard Stone, Jeffrey Mervis, Bruce Alberts, Elliot Marshall and Richard Kerr.

Thank you Pallab Ghosh, Vesa Niinikangas, Natasha Mitchell, Kathryn O'Hara, Lucy Calderon, Curtis Brainard, Dominique Forget, Chul Joong Kim, Christophe Mvondo and Jean-Marc Fleury, for facilitating insights into the global stage of science journalism.

We would also like to thank James Cornell and Boyce Rensberger and Deborah Blum. Our thanks to Raj Kamal Jha, Chief Editor of *The Indian Express* and the one with the magic touch, Mahendra Vyas, K Manohar, Mahesh Rangarajan, P Sunderarajan, friends who have stood by and bridged the distance between professionalism and personal warmth and affection. To fellow Martian scribes, Ajay Lele, Srinivas Laxman and TS Subramanian, our thanks are due.

A very special thank you to Suresh Gopal, Jyoti Mehrotra, Raj Bilochan Prasad and Kam Studio at Bloomsbury India, for turning this tight rope into a cakewalk!

Thanks are also due to RK Tuli and Bhavna Upneja. To friends who have always made us feel special – Saguna Dewan and Ravinder Singh Suri, Sunil and Kshama Gatade, Vijay and Deepa Naik, Kalpana Swamy, Padmanabha Rao, Anasuya Rao, Rakesh Naggar – warmest thanks.

Our work is meaningless without our family and it is hard to find the right words to thank them. To Nayantara and Ashwat Bagla, Sharad Bagla and Padma Menon, Chandralekha Varma, Gunjan and Smita Bagla, Aviral and Anshika, M Venugopal and Madhavi Menon, Mandakini Menon, Siddhartha Menon and Radhika Ranawat Menon, Harsh and Prerna Rastogi, Shraddha and Aastha, Niven and Sunanda Chaturvedi, Aditi and Arushi Chaturvedi, Amita, Dhruva and Bharath Kalyanram, Arun and Anil Bagla. To Sushma Kerketa, Maninder Singh, Susheela Andrews (Mary) our very special thanks. To Shershah Bagla (Sheru) without whose unquestioning love, no book gets written.

In closing, it is important to share that a lot of the deep motivation and drive to write this book emerged from having absorbed the inspiring and moving stories of Vikram Sarabhai's leadership at a time in the history of India when the only tools in hand were a handful of sky and a million ideas. That said, the blue sky is yours, and all the errors are ours!

Where science and Hinduism meet. Shiva's Nataraja pose which is considered as the cosmic dance of the Universe called the Tandava. This giant Chola bronze was gifted by the Government of India to CERN, the European Nuclear Research Centre in Geneva, where the plaque says: 'O omnipresent, the embodiment of all virtues, the creator of this cosmic Universe, the king of dancers, who dances the Ananda Tandava in the twilight, I salute thee' – Sivanandalahari, by Sri Adi Sankara.

Credit: Pallava Bagla

Chapter 1

The Big Picture of India's Space Ambitions

In a slick and well-appointed room – reminiscent of science fiction movies – where flashing banks of computer screens are dominant, with large plasma displays gleaming off the front wall, a final countdown has begun, in a firm, clear voice that has inflections of a south Indian accent. Just moments before, there are other voices of people who are ticking the boxes of actions that remain to be taken before the towering, rugged rocket blasts off into the clouds. The setting is surreal, a sci-fi like structure in a rural landscape. The Mission Control Centre at the Satish Dhawan Space Centre (SDSC) on the otherwise quite unexceptional barrier island of Sriharikota off the Coromandel coast in the Bay of Bengal resembles a flying saucer. Inside the control room, the tension is palpable, but only in an under-stated sort of way. In the minds of space scientists and engineers occupying the room there is the cold and clinical analysis of severe concentration on standard protocol, but there is also whispered and fervent prayer – for the launch to proceed without a glitch. A 56-hour countdown began at 6 a.m. on November 3, 2013, methodically ticking a list of 803 different actions that need to be checked before the final nod for the launch. With just minutes left for the launch, the last 100 or so actions are verified with precision, and the Polar Satellite Launch Vehicle of the C 25 version, lifts off, piercing the clouds for the twenty-fifth time in its history, beginning one dramatic and long journey – to the Earth's planetary neighbour, the God of War, Mars. The island, desolate till not so long ago, is abuzz.

Earlier in the day, four balloons float into the sunny and clear skies, checking wind conditions up to a height of 25 kilometres. The balloons bring back data of a fine day. What a relief for those watching the skies – since the month of November is notorious along the eastern coast of India where the weather can get tricky. What if any adverse atmospheric event occurs, what would the mission scientists do? November happened to be the month of choice because of the relative positioning of orbits of the Earth and Mars, being the most comfortable configuration possible in the unforgiving environment of space and the circumstances of an inter-planetary journey. This is the first time ever that any space launch is taking place in November from the SDSC.

The media centre is crawling with journalists from the world's media, descending on the high-tech space port at Sriharikota to cover India's maiden inter-planetary mission – news that will surely make history. Just a few minutes before the launch, a dark cloud appears over the island, bringing with it a sudden burst of rain and the sighs of sinking hearts. Will they call it off, many ask. But there is no lightening, the real adversary of rocket launches – so that's a big relief. The launch proceeds as planned and excitement spreads rapidly. Back at Mission Control, the plasma screens sparkle with the flames of the rocket's ignition at its tail-end and a small flank of rocket scientists gazes at their creation lifting off. Quick relief and satisfaction spreads but palpable anxiety remains, as they continue their observations of the huge vehicle's trajectory. But there is no doubt at all that India's journey from the Red Fort to the red planet has begun.

Two days in the month of November! Two destinations, two rockets – one that goes up to check out the Earth's upper atmosphere to see what it is like, and the other, scaling the giddy heights of ambition and defying the rules – for countries like India are not supposed to harbour crazy dreams about romping around in space – to lug a heavy satellite that is to travel all the way to Mars. A span of 50 years separates these two events between 1963 and 2013, and it doesn't take much to gauge that these have been dramatic years of action for tens of thousands of people who have worked for India's space programme. These are years that have brought India to a watershed moment; actually setting out to Mars with a small 1,337 kgs satellite – the size and weight of a Nano car – wrapped in a golden thermal blanket, nestled within the top encasing of the rocket. From 1963, the space programme has trundled through half a century, moving from the lovely, secluded beach sands of the Arabian Sea at Thumba, a tiny hamlet near Thiruvananthapuram's Veli Hills, to the futuristic and Star Wars like Mission Operations Complex (MOX) at the ISRO Telemetry, Tracking and Command (ISTRAC) in Bangalore, with exotic locales for ground stations in Port Blair on the Andaman and Nicobar Islands, Brunei, Biak in Indonesia and Mauritius, and also in Bangalore, Lucknow, Sriharikota and Thiruvananthapuram, and two ships, Nalanda and Yamuna, in the south Pacific Ocean. It is for the first time that a ship-based tracking platform is being used for a mission. MOX, with its plush seats and working stations laid out in a crescent-moon formation around large, high fidelity screens filled with blips and animations of the latest missions, feels light years away

The Polar Satellite Launch Vehicle, standing tall, almost as high as a 15-storey building and the weight of about 50 full-grown elephants, in its 25th flight put India's Mangalyaan in space.

Credit: ISRO

The sleepy village of Thumba in Kerala in 1963 witnessed the first-ever launch of a modern rocket from India. The American made Nike Apache rocket seen in the photo was launched on November 21, 1963 at 6.25 p.m. and rose a glorious 180 km into the sky.

Credit: ISRO

from the tiny fishing community of Thumba. This tiny dot on Kerala's map was distinctive and special to Vikram Sarabhai, the physicist-turned-space scientist who fathered India's space programme, because the geo-magnetic equator passes through it. Procuring the Nike Apache, a small, sounding rocket, from the United States of America, Sarabhai and his colleagues made a grand show of its launch, doing so to assess physical parameters of the upper atmosphere, but also to announce to the world that a nation just a decade and a bit into its well-earned freedom from the British Raj is ready to rub shoulders with a small clutch of giant nations in the world that have made the Universe their playground. The action at

Modern thinkers who saw the power of technology. In this composite image displayed proudly by ISRO are (L to R) – Vikram Sarabhai, the Father of India's space programme; Homi J Bhabha, the Father of India's nuclear programme; and, Indira Gandhi, the then Prime Minister of India.

Credit: ISRO

Thumba was humble, to say the least, but the rocket went up a glorious 180 kilometres into the atmosphere. The Kerala Legislative Assembly that was then in session, took a break for a few minutes to step out and watch the spectacle. Today, India is in a select club of top space-faring nations, along with the former USSR, the US, France, China, Japan, and the European Space Agency (ESA).

The south of India is pleasant in November, unlike the north that begins to feel a strong sense of an oncoming winter, but then that's thousands of kilometres away. It is a particularly balmy afternoon on Tuesday, November 5, 2013, when India's USD 70 million (₹ 4,500 million) Mars Orbiter Mission or Mangalyaan (*mangal* means a lot of things in Hindi – Mars, Tuesday, and anything auspicious) takes off, with an unmanned satellite built by about 500 scientists, beginning a 680 million-kilometre journey of 300 days to rendezvous with the red planet, making India part of a small group that includes Japan, China, Russia, the US, and the ESA, that have attempted space travel to Mars. Of these, only the latter three have succeeded. Since 1960, some 51 missions have been launched, about a third of which have ended in disaster, the most recent being the Chinese failure in 2011. If India does make it to Mars, it would really only be the third nation in the world to have done it all on its own after the US and Russia. The ESA has also been successful.

A million questions arise in a million minds across the world – how come India wants to go to Mars? Needless opulence is how many perceive the space programme, and in the need hierarchy of human beings with food and shelter on top, it might seem truly unnecessary. The debate is valid, and starkly divided – it is either a giant leap or a fool hardy step by a nation that still cannot provide electricity to 400 million people of its population, or save babies from malnutrition, or ensure toilets and sanitation for all. It all depends on which side of the divide you belong, but that doesn't change the fact that in the big picture of human development and nation-building, it is not an either-or question at all. If anything, the Universe is non-discriminating and all-encompassing, it is ours to discover, regardless of which nation we belong to. Humans everywhere are the same, and deeply attracted by the call of the frontier, so to speak. To gaze at and imagine starry encounters, to seek answers to earthly puzzles, and to interminably feed the fire of human curiosity. In its vastness of many hues, of lightness and darkness, there are infinite secrets, intimately linked to infinite questions that the human brain keeps asking. In the billions of years that the Universe has been around, it was the philosophers of ancient Greece, centuries ago, who formed the first questions to seek answers. From then, right into the twenty-first century, there has been no looking back. Questions, imagination, questions again, telescopes, more questions and then the first

The headquarters of the Indian Space Research Organisation (ISRO) in Bangalore with a bust of Vikram Sarabhai prominent in the foreground.
Credit: Pallava Bagla

Early humble beginnings: The nose cone of an Indian rocket being carried on a cycle.
Credit: ISRO

Hard-working scientists nurture the Indian rocket programme; an ISRO engineer transporting a rocket on his bicycle.
Credit: ISRO

baby steps outward, finally leap-frogging into advanced space exploration programmes – the world has been busy thinking about and discovering space. Baffling mysteries and ineffable romanticism are what surround these explorations, bringing to the study and exploration of space science and technology an irresistibility for most people, and therefore to most nations. In fact, if it wasn't for this enigmatic attraction, we would not have possessed even a fraction of the information and knowledge base we do today as a global society made unique by this very thirst – to know, and to use knowledge for action. Much of this knowledge is essential for human development and progress of society and civilisation, but much is also simply a fun-filled gathering of novel information, a childlike unearthing of facts that get strung together to tell charmed and captivating stories of unknown biological truths, uncharted territories and mysterious extra-terrestrial frontiers and beyond. India's space programme is also viewed by many as part of an Asian space race: while the twentieth century saw the US and the then Soviet Union racing to outdo each other in space, as two global superpowers, the twenty-first century has the two regional rivals China and India, doing the same thing. China took a leap and in 2003 sent off its first astronaut into space. India, in contrast, continues to take a lead in robotic missions. India's maiden mission to Mars has in many respects outshone China in inter-planetary missions.

But, let there be no misconception that the massive investments that nations make for studying outer space and planets are only to satiate curiosity. They do so because the atmosphere that

Humanity hopes to colonise Mars. An artist's impression of what a human colony on Mars would look like, capturing dramatically what millions of humans across the world dream of secretly. US President Barack Obama has mandated NASA to land humans on Mars as early as possible.

Credit: NASA

History retained: The first Indian satellite dishes used to gather remote sensing data at the National Remote Sensing Centre, Hyderabad.

Credit: Pallava Bagla

surrounds the Earth and outer space are now a basic human need, a resource that modern society cannot do without. Just a month before the Mars mission was launched, Cyclone Phailin made landfall on the same eastern coast of India, not far from where the Mars mission blasted off, but the deployment of modern technology and disaster warning systems saved thousands of lives, and the final death toll was lower than 20, with seven people dying in Odisha, the state that took the brunt of the storm. These were unbelievable figures, because less than 15 years ago, the same area had been devastated by a cyclone in 1999, and 10,000 people had lost their lives. Back in the 1970s, another Bay of Bengal cyclone took a toll of 3,00,000 lives. It is no coincidence that, in 2013, India had a fleet of eight weather monitoring satellites, and during Phailin's most crucial days of landfall, more than one hundred pictures of the cyclone were provided by Indian satellites and used by the India Meteorological Department, and both national and local disaster management authorities. Timely and accurate advance information about the cyclone hitting the coast saved thousands of lives.

Fishermen in many parts of India take the help of advisories facilitated by the Department of Space to locate the best fishing zones on the high seas, and it is the OceanSat satellite that aids them. India's forest zones are mapped using satellites, and areas most vulnerable for deforestation are identified from the sky, helping the conservation of wild biological resources. Getting pride of place as the information technology hub of the world with a booming software industry has also been because of the availability of communication satellites connecting real time Bangalore with Boston and all other such locations. Services for security forces that protect nations' boundaries, mapping the globe, locating natural resources, climate and weather, transport and navigation, GPS, disaster management – all this and more is because satellites have been up there, doing their job as directed from the slick and snazzy ground stations run by armies of engineers, technicians and scientists. We would not have television and Direct-to-Home services if it wasn't for satellites in space, nor would we have the explosion of telephony that people enjoy today. The country's vibrant media

A replica of India's first small rocket from the Rohini series to be launched from Thumba.

Credit: Pallava Bagla

India gate-crashed into the global space age, riding on an unlikely vehicle of progress, the humble bullock cart. Seen at the ISRO Satellite Centre in Peenya, near Bangalore is the APPLE satellite, riding on the cart for testing in a non-magnetic environment.

Credit: ISRO

thrives and survives on satellite-based linkages called electronic news gathering. The television and broadcasting boom – India's communication satellites have helped spawn more than 500 entertainment and private news television channels – is breaking the literacy barrier and taking knowledge everywhere in the world's largest democracy with a population of 1.21 billion people. The demand for space-based transponders is so high that ISRO is unable to meet requirements and is forced to hire space from foreign vendors.

Of course, if it wasn't for space exploration, we also would not have known much about what surrounds us, what our neighbourhood is like, and what the big picture is in which the Earth moves around. It is obvious that the application of space for human development is what draws financial and governmental commitments for space programmes, not just the drive to unravel its magical mysteries, although it must be said that the two are inextricably linked – and that is critical for India. It is fascinating that current leadership of the space programme still follows, in complete earnest, the *mantra* used by Professor Sarabhai and Professor Satish Dhawan, his

Many uses of space. Scientists at the India Meteorological Department using satellite data to track a cyclone.

Credit: Pallava Bagla

immediate successor, supported as they were by political leadership and top decision-makers in the country. Priorities have remained steadfast – ISRO today focuses on helping large sections of the Indian people get access to services supported through its nifty space technology and is considered a global leader in space-based applications. Sarabhai's famous words, quoted most often, are intensely significant even decades after they were uttered: 'There are some who question the relevance of space activities in a developing nation. To us, there is no ambiguity of purpose. We do not have the fantasy of competing with the economically

Yuri Gagarin: The man who spurred generations of space explorers. Gagarin was a Russian cosmonaut and the first human being to fly into space on April 12, 1961, when he spent less than two hours orbiting the Earth.

Credit: Pallava Bagla

The small fishing village of Thumba has now become the hub of India's rocketry programme. The highly guarded Vikram Sarabhai Space Centre in Thiruvananthapuram.

Credit: Pallava Bagla

advanced nations in the exploration of the Moon or the planets or manned space-flight. But we are convinced that if we are to play a meaningful role nationally, and in the comity of nations, we must be second to none in the application of advanced technologies to the real problems of man and society.' Narendra Modi, while witnessing a satellite launch in June, 2014, just a month and a few days after being sworn in as India's fifteenth Prime Minister, echoed the same sentiment, but in the language of today: 'Many misunderstand space technology to be for the elite. That it has nothing to do with the common man. I however believe, such technology is fundamentally connected with the common man. As a change agent, it can empower and connect, to transform his life. Technology opens up new opportunities of development. And gives us new ways of addressing our challenges.' The Indian space programme was born of such thinking, and in just half a century, is in the midst of its first inter-planetary mission to the God of War, the ancient Roman mythological interpretation of Mars. Of course, there is no war-like frenzy on the sprawling campus of the Vikram Sarabhai Space Centre (VSSC) in the sliver-like deep southern state of Kerala, as ordinary men and women go about their work amidst higgledy-piggledy wilderness filled with lush, tropical vegetation characteristic of the equatorial regions of the world.

This is the place where the Indian space programme was born, made distinctive by the trail-blazing and charismatic Sarabhai who went there in search of an ideal spot from where to launch rockets. Thumba was a small fishing community of people, but it got carved on to the world map – as did its local parish, the Saint Mary Magdalene Church – because of the spectacular space programme that grew around it ever since Sarabhai dialogued with the Bishop and the community of the parish more than five decades ago. References to this church abound, because it was the only decent building in the area where some work could be done in a sheltered environment. Somehow, it is easier to conjure up the image of a small and modest building, meant for prayers of a small and modest fishing community. In reality, the Thumba church is a relatively large and impressive building, beautifully built and quite striking indeed. It was the first building where Indian scientists took their initial, tentative steps towards the Indian rocketry programme, and today houses a space science museum. It must be said that this is perhaps the only church in the world that has been transformed into an educational museum, while keeping the altar intact. Sarabhai, attending Sunday morning service before broaching the subject with the Bishop, must have thought through on what he would say. Soon after this exchange, the Bishop led his little community to cooperate with Sarabhai and accommodate his request, obviously another visionary who appreciated the importance of allowing science into his world of religion and prayerful practice. There is an odd consonance between these early events that drew connections with religion through the Thumba Bishop's imprint, and ISRO's current faith-based practices, subtle as they are. It seems that the Indian

The Mary Magdalene Church in Thumba, Thiruvananthapuram, the birthplace of the Indian space programme, where among others, President APJ Abdul Kalam cut his teeth making rockets. Now this place is a glorious museum dedicated to the Indian space programme.
Credit: ISRO

Inset: Classic stained glass windows of the Mary Magdalene Church, depicting the Solar System.
Credit: Pallava Bagla

space programme has always had some tenuous connections with religion! Whether it is the prayers that the chief of ISRO offers at Tirupathi before a major launch, or the church-turned-museum at Thumba where visitors to the museum quite involuntarily fold their hands and lower their heads in prayer as they walk past the old altar, or the *haldi-kumkum* discreetly marked onto several of the big pieces of equipment (a Hindu way of showing reverence for an object) across ISRO's many institutions, the connections show.

Catalysed by Sarabhai's chiselled diligence and his remarkable successors – Dhawan, U R Rao, K Kasturirangan, G Madhavan Nair, and K Radhakrishnan – the Indian space programme is today robust, bustling, and standing shoulder to shoulder with the programmes of the world's most developed nations (see Annexure 1). The Nike Apache was carried to the beach just behind the church in Thumba and fired into the sky, an accomplishment viewed by a glittering audience of India's top political leadership and many other global personalities. A young APJ Abdul Kalam was also part of the team, playing a critical role in the launch. This was India's missile man, to become famous for his work on ballistic missiles and launch

K Kasturirangan (R), then Chairman, ISRO, taking former Prime Minister PV Narasimha Rao around.

Credit: ISRO

vehicle technology and who, nearly four decades after the big bash at Thumba, becomes the eleventh President of India between 2002 and 2007. The successful launch of the Nike Apache was dramatic in more ways than one. In fact, the day after, when the Americans were congratulating the Indians on the launch, there was an interruption over their public address system – President John F Kennedy had just been assassinated. Meanwhile, the Polar Satellite Launch Vehicle (PSLV) completes its twenty-fourth consecutive successful job by transporting the Mars Orbiter Mission satellite, lovingly called MOM or Mangalyaan, onward into outer space on a long journey.

Book-ended by these two significant events is an exciting story that led India to develop what is rather seriously referred to as end-to-end capabilities in space in just half a century. From building its own heavy-duty rockets to fabricating sophisticated satellites and payloads, which are basically scientific instruments that are sent up on satellites to carry out specific scientific enquiry, perfecting the development of rugged avionics, smooth ground systems and countless other aspects of a holistic space programme, the ISRO folk have been kept busy. In fact, ISRO's hundredth mission in five decades took place on September 9, 2012, when Spot-6, a satellite from France, and Proiteres, a Japanese satellite made by students, went into space riding the C-21 version of the PSLV, from the Sriharikota launch pad. This was a history-making event. Inside the cavernous workshops and hangars at VSSC, rocket engineers and scientists potter about fussing over assemblies in different stages of development, and it all seems like a day's work, not history in the making. To Koshy M George, Deputy Director at VSSC, and a man who has spent a lifetime devoted to the Indian space programme, there's history, but there's also pride. The large exit doors make easy the imagery of how a finished rocket would trundle out of this campus, enormous rocket assemblies enroute to Sriharikota, carefully packed into large transport trucks that will travel slowly, very slowly. Till just some years ago, a lot of components and materials were imported, but George's pride is in indigenisation – India builds and launches its own heavy duty rockets. Right up till 2004, 90 per cent materials for rocket fabrication were imported, and now it's a mere 10 per cent, which is quite an achievement. Some 45 rockets have blasted off from the twin launch pads at Sriharikota. The PSLV, ISRO's versatile workhorse rocket, has an enviable record of more than two dozen consecutively successful launches. In its heaviest variant it weighs 320 ton – as much as a fully-loaded 747 Boeing Jumbo jet at lift off, and is as tall as a 15-story building (about 44 metre high), with the ability to carry 1.5 ton to a geo-synchronous transfer orbit and about 3 ton to a low Earth orbit. It's the same workhorse that was used for the MOM.

Developing indigenous satellites: ISRO's Satellite Centre in Bangalore.

Credit: Pallava Bagla

Communicating in deep space: ISRO's deep space network at Byalalu, outside Bangalore.

Credit: Pallava Bagla

Cold war in deep space. India and the United States of America duelled strongly for decades as the former attempted to buy the sophisticated cryogenic engine technology from Russia.

Credit: Pallava Bagla

The PSLV launcher has been commercially used by the Italians, the Israelis and the French for their satellites. In all, India has launched a total of 40 foreign satellites from its soil, and with a major drive to increase indigenisation, the last decade has seen more than ₹ 5,000 million worth of infrastructure being built to help indigenise production. Dr P Kunhikrishnan, Project Director for the PSLV, believes that getting most of the work by industry will be a major achievement, enabling them to focus on newer and more creative horizons of discovery. Today, the attention is on getting the next generation launch vehicle, the Geosynchronous Satellite Launch Vehicle (GSLV) Mark III successfully off the ground. It will allow heavier satellites in the 4–5 ton class to be transported into space orbits, and have much more power. Aptly, the GSLV Mk II has been called the naughty boy of the rocket family at ISRO. Mastering the art of flying India's heavier rocket that can hoist the 2.5 ton class of communication satellites into orbits has had a chequered history. The 49 metre, 402 ton rocket was first tested on April 18, 2001, and since then eight flights have taken place of which five have failed. The rocket programme suffered a crippling blow when in 2010, GSLV had dramatic twin back-to-back failures. ISRO redeemed its lost glory when in January 2014, it successfully launched a fully-indigenous version GSLV Mk II from Sriharikota. This was an event of geo-political significance. In the early 1990s when India embarked on the first development cycle of this rocket, the country needed a highly sophisticated rocket motor called the cryogenic engine, to launch large and heavy communications satellites. But, after India exploded a nuclear bomb at Pokharan, a stifling technology denial regime led the US to pressurise the Russians, who then back-tracked on a deal to sell to India the know-how for making cryogenic engines. Instead, they provided seven ready-made cryogenic engines that could be used by India. This restriction was a severe setback for India's space ambitions, but when denied fish for dinner, you just go out and learn to fish. And so it was that after a delay of almost two decades, the GSLV has been launched successfully. This launch has not only thwarted the very purpose of sanctions being imposed on the country, but also paved the way for the GSLV Mk III *avatar* which might become the preferred vehicle for sending humans into space from Indian soil.

ISRO also designs and fabricates some of the most sophisticated satellites, and with ten currently in orbit, India has the largest fleet of communication satellites in space among all countries

The taming of the naughty boy; ISRO glitterati of the day celebrating the successful launch of the GSLV Mk II on January 5, 2014. L to R: MYS Prasad, Director, Satish Dhawan Space Centre, M Chandradathan, Director, Liquid Propulsion Systems Centre, S Ramakrishnan, Director, VSSC, K Sivan, Project Director, GSLV, NR Vishnu Kartha, Project Director, Cryo Upper Stage Project, K Radhakrishnan, Chairman, ISRO, M Nageswara Rao, Project Director, GSAT-14, AS Kiran Kumar, Director, SAC, SK Shivakumar, Director, ISAC and VK Dadhwal, Director, National Remote Sensing Centre.

Credit: ISRO

in the Asia-Pacific region. In 2009, India's maiden mission to the Moon, Chandrayaan-1, brought back the first clinching evidence of the presence of water on the parched lunar surface. In more down-to-earth missions, Indian satellites are also helping locate underground water aquifers for some of the poorest and most marginalised people of India. Flying as many as a dozen remote sensing satellites, New Delhi controls the largest fleet of civilian eyes in the sky in the entire world, with some birds that can see objects as small as a car from 800 kilometres in the sky, with other satellites that have day and night viewing capability and can image any part of the world. These are birds that ought to send a chill down the spines of India's enemies as they are some of the best spy satellites in the world. Till recently, the United States Department of Agriculture was estimating crop yields on American farms using data sourced from India's 'ResourceSat' satellite. What a contrast from the time when India used to import images from American LandSat satellites to try and forecast famines! Aryabhata, the first Indian satellite, was launched on April 19, 1975. Quite incredibly it seems, while Sarabhai had envisioned an Indian society that would be serviced in many ways through space technology and applications, today there is a global dimension to that dream.

Students at work: Jugnu (meaning 'firefly' in Hindi), a tiny satellite made by the Indian Institute of Technology, Kanpur for mapping vegetation. It was launched on October 12, 2011 from Sriharikota using the PSLV rocket.

Credit: Pallava Bagla

Space icons Indians admire. L to R: G Madhavan Nair, former Chairman, ISRO; Sunita Williams, American astronaut of Indian origin; and, Rakesh Sharma, the first Indian astronaut in space.
Credit: Pallava Bagla

There is something to be said about the freedom to dream, and the power to make dreams turn into reality. When Dr K Radhakrishnan, Chairman of ISRO, leader of the space dreamers, was a young boy of seven he began learning Kathakali – Kerala's famous classical dance form that tells stories of Hindu mythology through dramatised performances and complicated costumes, elaborate make-up that takes around three hours of patient application in the case of a *purusha vesham* (male character). Kathakali brings to life stories of Hindu gods and goddesses, and various mythological characters, somewhere rather closely linked with the infinite spaces that form the Universe. And maybe somewhere in the mind of this engineer from a small township called Irinjalakuda in Kerala, just the way the mundane gives way to the ethereal when a Kathakali performer takes the stage and experiences a spiritual transformation as the mythological character he or she portrays, Radhakrishnan finds himself transported each time a space mission takes shape and finds success. The common denominator, really, is passion. Radhakrishnan is ardent when he talks about his work, his life fuel. But then the performing arts are a close second. Still a performer, Radhakrishnan can be found singing devotional, *Carnatic* music at the Ekadashi temple festivals in Guruvayoor, Kerala's world-famous temple dedicated to Krishna. Graduating from the Government Engineering College in Thrissur, this unassuming man began his journey in the space programme quietly enough, and now handles the top job. The art he practices is what gives him the spiritual courage to keep on regardless of the pressures of his work. These are strange connections; and one of the main buildings at VSSC has in its foyer a large portrait of Sarabhai in his signature white traditional Indian outfit of a *kurta-churidar*, and close to him, standing like a sentinel, is a large and beautiful model of a Kathakali dancer.

In India, it is commonplace for government organisations and institutions to be under-performers. Funded fully by public money, many are seen as large and labyrinthine, slothful and inefficient, affected by the lenient policies that govern human resource management, an unusual lack of accountability for performance, and a certain sense of security that is completely non-existent in the private sector. ISRO is the government sector's dark horse that

India's newly elected Prime Minister Narendra Modi in the centre, witnessed the twenty-seventh launch of the Polar Satellite Launch Vehicle (PSLV) on June 30, 2014. L to R: P Kunhikrishnan, Mission Director, PSLV; Chandrababu Naidu, Chief Minister of Andhra Pradesh; ESL Narasimhan, Governor, Andhra Pradesh; K Radhakrishnan, Chairman, ISRO; Prime Minister Narendra Modi; M Venkaiah Naidu, Union Minister for Urban Development and Parliamentary Affairs; Jitendra Singh, Union Minister for Science and Technology; MYS Prasad, Director, Satish Dhawan Space Centre, Sriharikota; Seshagiri Rao, Associate Director, Satish Dhawan Space Centre, Sriharikota.
Credit: Press Information Bureau, Government of India

upsets each of these definitions. The paradox is that the general human resource management policies, salaries and other such characteristics are not unique here, except that the Sixth Pay Commission of the Government of India brought good news for the staff, a special increment meant as a reward for the organisation's high performance. PM Modi has said: 'India's space program is a perfect example of my vision of Scale, Speed and Skill. Our space scientists have made us global leaders, in one of the most complex areas of modern technology. This shows that we can be the best. If we apply ourselves, we can meet the aspirations of our people.' In fact, when the MOM mission took off, Dr Manmohan Singh, the then Prime Minister, said while congratulating ISRO: 'You have risen to the occasion' and has also often been known to say that the best from ISRO is yet to come. The organisation, set up on India's Independence Day in 1969, is fuelled by an annual investment of just about USD 1 billion [the US's National Aeronautics and Space Administration (NASA) spends 17 times more, on an average], and is the custodian of all space technology for the country. The MOM mission is also in the spotlight because it is a highly cost-effective mission, costing each Indian just ₹4 as Radhakrishnan is fond of saying. The pride therefore, seems justified. Among the countless news reports that flooded the world at the time, a certain correspondent for a world-famous television news agency started her story by saying, 'India brags about the launch'. Indeed, it must seem puzzling that a country with seemingly unsurmountable developmental problems is scaling such heights in the exploration of space. Strangely though, coming up close with scientists and space techies who work for ISRO, one thing stands out – they never brag! In this organisation of doers where 16,000 people are employed, work is worship. Little else matters.

Whether at VSSC in its picturesque setting, the business-like Space Applications Centre (SAC) in Sarabhai's home town of Ahmedabad, the sprawling headquarters in the heart of Bangalore, or at the ranch-like Byalalu off Bangalore with its landscape reminiscent of the iconic Indian film *Sholay*, where giant antennas watch over spatial action, there seems to be a code for human behaviour in operation. Meet a young engineer in his early twenties, or speak with a senior official who has spent close to four decades with the institution, everybody is energetic, the body language is positive, the talk is generally peppy and animated, and one feels as though the best job in the world is to be part of the Indian space community. In an infinitely-repeated refrain, many ISRO staffers say that their founders set the tone, building a work culture based on openness and transparency, where each person is respected for whatever role he or she plays. People-centric policies that allow career pathways and progression, meritorious promotions and reward systems that are a speciality of the organisation, and team work and supportive supervision are keys to success. In fact, 'team work' takes on a very different connotation for those who work at ISRO. Every person has a minimum of two roles to play, one is his or her

Satellites in the service of society. These Idu Mishmi tribals experienced the information superhighway even before they were connected with the rest of the world through an all-weather road, given their remote location in the district headquarters of Anini in the Debang Valley district of Arunachal Pradesh, India's extreme northeastern corner. Till recently, this township with its small population of 2,500 people received all its supplies through a helicopter service. A major road development scheme is still underway to connect this strategic border town to the rest of India.

Credit: Pallava Bagla

core job, and the other is to be assigned an additional title, temporarily, as part of a project team set up for a specific, time-bound activity. Of course, the project role of a person is linked to his or her core skills and experience. What this means in the long-term is that people across disciplines, across geographies and across institutions, end up connecting with each other and working together at some point of time or the other.

There are basic professional reasons why ISRO is considered a people-centric and learning institution. Each activity or every project is controlled through stringent and robust technical review processes that are transparent and non-heirarchical, where every voice is heard as long as it is scientifically sound. Working systems are not altered unnecessarily and schedules, once fixed, are rarely tinkered with. Of course, in mastering complex technological design, time over-runs do happen. Then, when a person sees a concrete result and visibility of things happening, that's a great motivator too. Even the cooks and drivers who work for the organisation feel

India's abiding love affair with space exploration. Top Indian leadership at the Space Applications Centre in Ahmedabad. Former President APJ Abdul Kalam (centre) and current Prime Minister Narendra Modi (extreme right).

Credit: ISRO

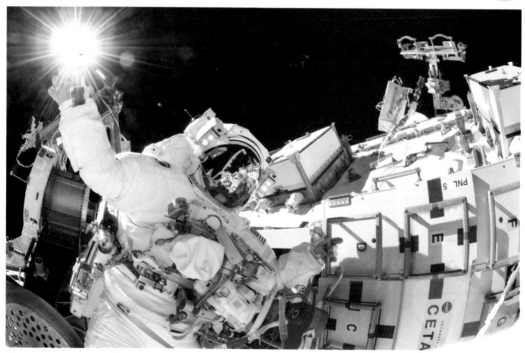

Astronaut Sunita Williams conducting a spacewalk. She gained the distinction of holding a world record by conducting space walks totalling 50 hours and 40 minutes among women.
Credit: NASA

like they have a part to play when a magnificent rocket tears into the clouds, leaving behind a plume of fire and smoke. The founding fathers of ISRO deliberately headquartered ISRO in the science city of Bangalore. Being more than 2,000 kilometres away from the national capital protected the organisation from needless political interference. The gamble has indeed paid off since the Department of Space (DOS) has mostly remained insulated from the unsavoury politics that is reminiscent of Lutyen's Delhi.

But most of all, it is about organisational focus and a clear vision of what needs to be done, by when, and how. The same clarity is communicated to decision-makers and policy-makers in the governmental hierarchy, creating a kind of shared vision between the government and the space scientists, and also cementing key programme priorities so that a back-and-forth is avoided, making clearances easier and financial allocations trouble-free. Needless to say, a lot of things make it easier for the space programme; since ISRO was born, it is the Prime Minister who is in charge of the DOS portfolio, and space programmes are also attended to with special care by national governments and the powers that be, given their close linkages with strategic affairs, defence, and national pride. Those who fuel the programme, with financial resources, like the Cabinet or the Planning Commission, also recognise the importance of this work and that it needs to be done. Radhakrishnan also believes that once the organisation commits itself to a particular project, and the money is made available, they stick to it doggedly. This endears them a lot to the policy makers.

Many of the senior leaders at ISRO now are engineers who joined as young chaps (yes, mainly men then) to work with Kalam and still recall him with great devotion and affection. 'Motivated, magical leadership' was his style, says one. And one good leader surely begets another. Committing themselves to 24 × 7 schedules, holding off enthusiastic parents and postponing matrimony, working for small salaries but big dreams and just knowing that they were beavering away at something that would make a difference – these are the stories heard on the campuses of ISRO's institutions. These are stories of young people; ISRO has 4,500 young staff today, with anywhere between one to 15 years of experience. These are also stories of senior staff members who have devoted lifetimes to their work, but have also got handsome returns. They get to continue higher studies if they so desire, with many moving to a mid-career MBA at India's top Indian Institutes of Management (IIMs), also obtaining doctorate degrees from prestigious institutions of higher learning like the Indian Institutes of Technology (IITs) and the Indian Institute of Science, Bangalore (IISc). It amuses Radhakrishnan how, with the success of the space programme over the years, his face has started appearing on advertisements that IIM-Bangalore places, highlighting distinguished alumni like him. 'I became a distinguished alumnus when Chandrayaan happened', he says, sounding pleased with the attention. These are all human stories that denote success, in more ways than one. Success that is not ethereal or indefinable like the Universe these men and women study, but success that is hard and real leading from the demonstration of utility and application, of an organisation that is a primary vehicle for India's development.

India's maiden mission to the Moon, Chandrayaan-1, made history on several counts. It discovered the presence of water molecules on the parched surface of the Moon. In a rare distinction, the prestigious American journal Science displayed that scientific finding as a cover on its October 23, 2009 issue.

Credit: Science

But it is impossible to contextualise everything about space science and technology into neat little boxes of service provision. That is because the adrenaline of scientific discovery is almost as essential as the physiological essentials, like blood and oxygen and beating hearts, but in an inexplicable sort of way. Moreover, since it is scientific discovery that adds invaluable dimensions to human development and the necessities and comforts of contemporary life, there really is no 'ambiguity of purpose' in the pursuance of space exploration, as Sarabhai famously remarked in defence of why a country like India needed to get into such seemingly glamorous activities. This quest of new knowledge is a characteristic of Indian society; in fact India is among the original cradles of human civilisation that has continued over thousands and thousands of years. In a strangely paradoxical way, this ancient nature of Indian society has seen reflections of itself in the ultra-modern and cutting-edge activities that characterise India's highly action-packed, super-efficient space programme, which has been coursing new

India made history on November 14, 2008 when the country's Moon Impact Probe (MIP) carrying India's flag was sent to the lunar surface. This is a replica of the plaque that was carried on board Chandrayaan-1, but with autographs of the key personnel involved in India's maiden mission to the Moon. Clockwise from above, the signatures are of: M Annadurai, then Project Director of Chandrayaan-1; Narendra Bhandari, the first Indian lunar scientist to propose a mission to the Moon; JN Goswami, then Chief Scientist for the Moon mission; George Koshy, then In-Charge, PSLV which carried the satellite into space; and, G MadhavanNair, then Chairman, ISRO who executed the Moon mission. This historical plaque is part of the personal collection of the authors.

Credit: Pallava Bagla

trajectories. It seems only natural that India has just embarked on its first deep-space mission to Mars.

There is a mystical allure that the red planet holds, perhaps mainly because it is the only planet that seems somewhat similar to Earth, urging us to often ask the question 'is there life on Mars?' Although really far away, Mars is after all, Earth's next-door neighbour in the planetary system. The red planet also has seasons, solid surfaces, polar ice caps, a 24-hour 39-minute day, perhaps water. The questions are many, and the Indian space programme holds the global distinction in being one of the very few countries in the world to attempt the deep space challenge of trying to get to Mars before all these questions can be answered. The ISRO scientists had been grappling with the idea to go to Mars from around 2007, but the actual

MOM sees Mother Earth! On November 19, 2013, even as Mangalyaan was racing towards Mars engineers at ISRO trained the Mars Colour Camera towards the Earth, and it clicked a picture that was beamed back. This picture caught the imagination of countless Indians because it dramatically captured a developing cyclone Helen in the Bay of Bengal and also showed parts of India.
Credit: ISRO

development and implementation of the MOM happened in a record time of 15 months beginning with the official financial approvals from the Government of India in July 2012. November 2013 was an important date that offered a unique launch window, and the next may have occurred only in 2016. So, with this great urgency, the MOM satellite was sent off on November 5, 2013, carrying with it the Mars Orbiter satellite with five instruments on board, each meant for a specific purpose of capturing new knowledge and facts. And amazingly, this is an end-to-end indigenous mission. Dr M Annadurai's ever-smiling and cheerful countenance obviously doesn't make one think of an exacting task-master but as Programme Director of the MOM, he must have some special talents! The man behind Chandrayaan-1, Annadurai has the famous NASA phrase, 'Failure is not an option', placed on his table, and seems to live the words. Once MOM was conceptualised it was clear to Annadurai: 'We are going alone on this one.' This decision of flying solo took ISRO on a gruelling 18 month roller-coaster schedule, that would surely have often been lightened by his hearty laugh, five parallel teams focused on the launch vehicle, the satellite, the post-launch plan, ground systems and the development of all payloads. Diverse teams working discretely and with each other created the architectural marvel that is the rocket that has carried the MOM satellite, the construction of the satellite itself, the design and creation of the payloads that will make possible capturing all necessary information and are currently on the satellite, and the ground systems and networks that will work ceaselessly to control the journey. It was a now-or-never effort in more ways than one. In India there is no luxury of making mistakes, according to Annadurai. Sounds a bit hard to believe knowing India? That is exactly how different the space sector and its governance are from others: there can be one, fully focused effort, and it must be perfect. So the stakes are high.

'MOM views Mother India!' was the first caption used to detail some of the early pictures that were sent back by the Mars Orbiter as it began its journey. A dramatic picture of a severe cyclone called *Helen* that was building up in the Bay of Bengal was taken by the Mars Colour Camera (MCC) and made for great viewing. To Dr Ashutosh Arya, senior geologist and scientist at SAC in Ahmedabad and Principal Investigator for the MCC, this was the picture to make the right impact with, since taxpayers would like it and find answers to their expressed

The official logo of the Mars Orbiter Mission, cheekily suggesting that Mars is for men.
Credit: Pallava Bagla

or unexpressed questions about the MOM and its utility. A mission like MOM would be meaningless without the power of being able to ask some tantalising scientific questions and unearthing some unknown and exciting scientific facts. The excitement in the high-ceilinged, clean rooms at SAC is felt weeks and months after the small and creative instruments teams created here have flown out to Mars. Several men and women in whites walk around in the rooms, with sparkling eyes, excited voices and so much to say about their little babies that are nestled within the satellite, unimaginably far from Sarabhai's home town. The SAC team hit the jackpot on scientific instruments getting on-board the satellite, with three of the five payloads having been developed here. ISRO's Advisory Committee for Space Science (ADCOS), under the leadership of its former Chairman, Professor UR Rao, first short-listed 25 proposals submitted by different ISRO centres, and then finally short-listed nine in July 2012. The SAC team knew that this was a race against time – with the launch window being small and fixed, there was no question of delay or leeway for that matter. Somya Sarkar, Associate Project Director for payloads on the Mars mission, wears a wry expression when he says that a normal payload development cycle is four to five years, and suddenly, they were faced with a challenge – three payloads to be developed in less than two years, and to be kept extra-light and compact, consuming minimum power besides being rugged for the unforgiving spatial environment in which they would need to perform and do their job. So the usual custom-built route was abandoned and Indian *jugaad* came in handy. Three light-weight and powerful payloads were on the list – the MCC as earlier mentioned, the Thermal Infrared Imaging Spectrometer (TIS), and a Methane Sensor for Mars (MSM). Two more payloads were being designed – the Mars Exospheric Neutral Composition Analyser (MENCA) at VSSC and the Lyman Alpha Photometer in Bangalore. These tiny but powerful machines will be kept busy once the satellite starts moving around in the Martian orbit. They would be taking pictures, measuring temperature, looking for methane gas, studying geological and other characteristics of the planet, trying to capture areas covered by carbon-dioxide ice when viewing the poles and much more. The MCC of course, will remain like the eye of the entire mission. Put in simple language, these little scientifically-designed machines will be those which would fill some of the blanks in human knowledge about Earth's neighbour, and quench peoples' thirst of wanting to know more about its alluring and mysterious presence.

Speaking of Martian allure, within a month of the launch of the Indian Mars mission, a global private announcement by the not-for-profit Mars One Foundation invited applications from people interested in going on a one-way trip to Mars in 2023 which attracted 2,00,000 applicants, including 20,000 Indians! The organisation announced that it plans to ready the environment there for human habitation and hopes to send four people every two years after 2024. So the call of the unknown still excites, that is certain. Even as a record number

The Polar Satellite Launch Vehicle (PSLV) in its twenty-seventh launch on June 30, 2014 placed in orbit five foreign satellites. This fourth fully commercial launch of the PSLV was witnessed from Sriharikota by Prime Minister Narendra Modi, who after the launch asked ISRO to make a satellite system for the benefit of India's neighbours and exclaimed '*yeh dil maange more*'.

Credit: ISRO

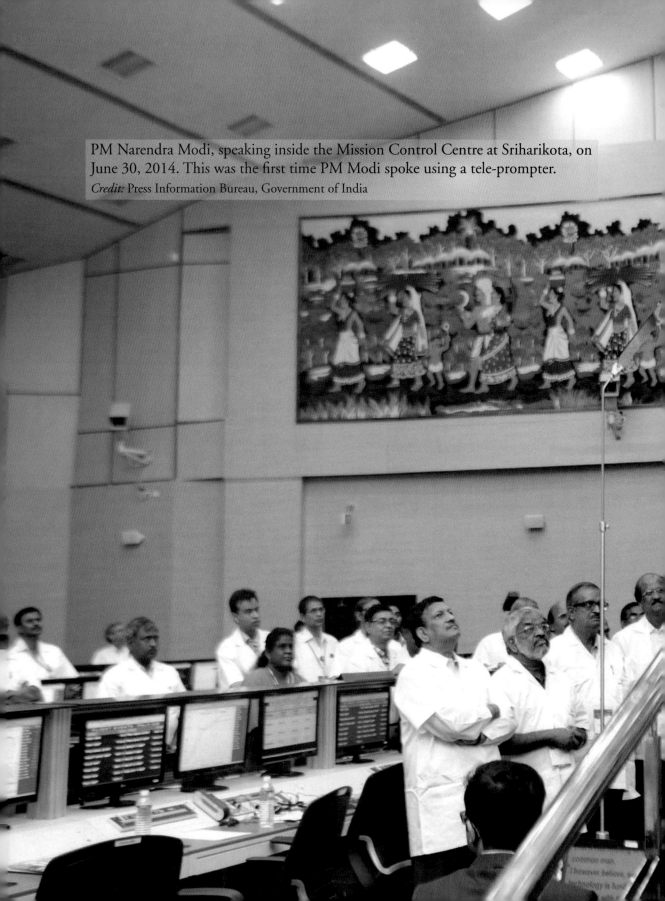

PM Narendra Modi, speaking inside the Mission Control Centre at Sriharikota, on June 30, 2014. This was the first time PM Modi spoke using a tele-prompter.

Credit: Press Information Bureau, Government of India

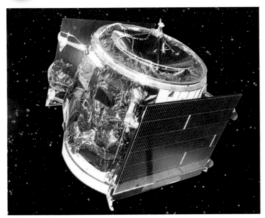

The Apple satellite: The Ariane Passenger Payload Experiment was an experimental communications satellite launched on June 19, 1981 using the Ariane 1 rocket. The satellite was used for relaying television and radio programmes.

Credit: ISRO

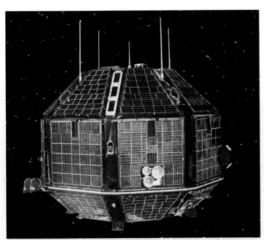

The Bhaskara satellite. These twin Bhaskara satellites were made by ISRO and were India's first Earth Observation Satellites. Bhaskara 1 weighed 444 kgs and was launched on June 7, 1979 from Russia. The on-board TV camera sent back images that were used in forestry and hydrology.

Credit: ISRO

of people put in their applications for this first unbelievably long deep space journey, to become the first settlers on a new planet, the Indian Mars mission Facebook page also attracts a huge number of visitors and 'likes'. Scientists managing this public engagement and outreach for ISRO say the social media response to this mission has been stupendous, revealing the innate scientific curiosity of the human mind and upping ISRO's popularity among the young by many levels.

Indeed, there are stories to tell. Stories of great work, facilitated by the laser focus and people-friendly policies of the Indian space programme, of getting deeply involved in the tantalising and irresistible tide of space exploration. Such is this story of India's journey to discover Mars, taking its first inter-planetary step by relying on these armies of extraordinary people. It is about feats of technological and scientific operation that is world class, but is equally a compelling, human narrative of unglamorous and unassuming men and women collectivised by and unerring discipline and highly effective human resource management principles that create such a work-conducive environment for ordinary engineering graduates that they invariably join ISRO as a first job and never ever leave, till they retire. Not just that, they learn never to let their motivation and interest and creativity sag. This is not chance, but a deliberate human resource management strategy that has at its fulcrum the philosophy of ultimate team work and the desire to succeed – the inter-dependency within different systems and groups (entities, as they are called in ISRO parlance) making team work the only practical *mantra*, with each one fuelled by a single driving force – that sending a satellite into space is either 100 per cent success or a total zero, a failure. There are no half-measures in space science and technology – whoever has heard of anybody claiming they managed to get the rocket half way up, or dropped a satellite off just before its planned orbit! It is fascinating that the majority of people who run the space programme are products of Indian colleges and universities, and the ISRO hall of fame is dominated by those with Indian degrees of higher education. Most of these people have studied engineering from regional engineering colleges, and very few are from the IITs.

It isn't like ISRO and the Indian space programme haven't had their dark moments. There have been many. The infamous S-Band controversy, also referred to as the Devas-Antrix deal that hit the country in 2010 is still not

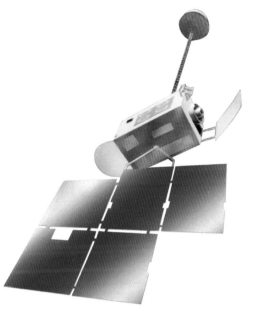

The first in the series of the Indian National Satellite System called INSAT, seen here as an artist's impression. This communications satellite, weighing about 1,100 kgs, were made by Ford Aerospace and were launched from the US.

Credit: ISRO

over. This was when the squeaky clean image of ISRO took a hit with alleged accusations that the esteemed organisation had connived with a private telecom company Devas Technologies to sell high value S-Band spectrum for a pittance to the same company. The commercial arm of ISRO, Antrix Corporation continues to be engaged in an arbitration in a bid to absolve ISRO and its leadership of short-selling scarce satellite bandwidth. In a similar vein, the space programme's image was seriously tarnished in 1994, when its scientists working on the crucial technology development programme for making an Indian cryogenic engine were accused of spying on behalf of foreign countries. Subsequently, the Supreme Court of India absolved all the scientists named in the case and found them 'not guilty'. But at the same time, technology development on cryogenic engines suffered a severe setback.

But the future remains bright, as always. Packed schedules with significant future projects are a constant at ISRO. Teams are focusing on deconstructing lessons from the country's maiden Moon mission and Chandrayaan-2 that hopes to land a rover on the Moon's surface is underway, as is a flight to study the Sun, Aditya, and possibly a visit to a passing asteroid. Finally, and most exciting – plans for sending humans into space. At the very young age of 22, S Unnikrishnan Nair joined ISRO after completing his BTech in mechanical engineering. In fact, while still studying for his last semester, he appeared for an interview and got selected. Today, the soft-spoken Malayalam literature buff is quietly beavering away at the Human Space Flight project at ISRO as its Project Director. Of course, there is not much to speak of right now, with a lot of the detail still under wraps, because the Indian government is yet to provide full approvals and clearances. There is however, enough seed money so that Indian space technologists can master the technologies needed for orbiting an astronaut in space. There are plans to use the maiden flight of the GSLV Mk III in 2014 to test the crew module and capsule that will carry the astronauts.

Today, at the Indian Institute of Space Technology (IIST) in Thiruvananthapuram, hundreds of young students are enrolled in a specialised space engineering and technology under-graduate course and get transformed into a ready resource for ISRO to select for jobs as soon as they complete their degrees. And the chances are stacked in favour of their staying on for decades, working, learning, progressing and also seeing the results of their toils. The Mars Mission, in a million ways, acknowledges the fact that science and scientific discovery, and the involvement of young people in that journey, is a reality in contemporary India. It challenges the cynical observation that young people don't care about the sciences today. They do, but only if it connects with them. That is perhaps why young staffers at ISRO are given their due. Like Nilanjan Routh, the 32-year old from Tripura, who looks after the Facebook effort at ISRO. As Shamsuddin, the 20-something scientist who is on the same team, puts it: 'More than 50 per cent of our audience is people between the ages of 18–30, and they encourage us to ask ourselves new questions, and we are pushed to explain scientific complexities in a way that is technically accurate but understandable to a layman.'

Chapter 2

Spurring India to Go to Mars: Marshalling the Mission

At the Lal Quila in Old Delhi I always feel overwhelmed. It is a fleeting experience of the confluence of ancient history with the bedlam of a contemporary, crowded mega-city. The cacophony of Chandni Chowk, one of the most historic and colourful bazaars of Asia, mixing with the ascending and impassioned notes of the *azaan*, the call for prayers from that other gem of history – the Jama Masjid – right across the road. The colourful environs of the Jain Bird Hospital, possibly the only one of its kind, add their unique stamp to the feel. The ramparts of the spectacular Lal Quila built by the Mughal emperor Shah Jahan and known as the Red Fort to the world, seem to watch over the infinite sea of carts and rickshaws, buses and bicycles in a meditative kind of way, indulgent of its incredible surroundings. From these very ramparts, economist Dr Manmohan Singh, former Prime Minister of India, announced with a hint of pride in his somewhat toneless voice that India was on its way to Mars, riding on the 'Mangalyaan'. That was on August 15, 2012, India's sixty-fifth Independence Day. Prime Ministerial speeches are considered significant because strategic policy commitments and priorities for the year are shared with the general public of the world's largest democracy. When I heard those words, I could feel the excitement coursing through my veins. What a story, I thought, already composing the opening lines in my head, of an inter-planetary mission following the rendezvous with the Moon. Now that is a story that would glide off the newspage. This was really a winner for a science communicator who lives for the news and for telling the story.

But the clam-up after that! I was not prepared for the silence that followed. It was like a persistent black-out of information on anything to do with the Mars Mission. And then, finally, came a different kind of 9/11. If one was a fiery destruction, the other was the pinnacle of productivity and human ingenuity. On September 11, 2013, ISRO finally relented and gave me unparalleled access to visit the clean room at the Satellite Centre (ISAC) where the MOM satellite was surrounded by scores of engineers and technicians in a feverish burst of activity. Not only did I get a ring-side view, but also had the unique and singular privilege of actually being allowed to touch the almost final configuration of the satellite. Wearing a white coat and cap, after walking through the gust of air from the dust-free purifier, it felt special. In all humility I can say that my finger prints are on a satellite headed to the red planet. For someone like me, who has a fascination for space, this is a dramatic thought and a rare moment to cherish through a life time.

Big science they say is fuelled by big dreams, but big dreams can become real only if they are dreamt early enough. This cannot be truer for any country than it is for India, where a crowd of urgent developmental priorities jostle and elbow for attention in the policy-making and action agendas of governance. And it is also in countries like India that the otherwise mundane question 'Why Mars?' resounds with very special meaning, triggering both complimentary and censorious debate and discourse. Some might say that it is two failures that led India to head to Mars: One was the twin failure of the GSLV in 2010, pushing the big rocket off the immediate horizon of ISRO's plans; the other, the somewhat unexpected failure of a Chinese Mars mission. On November 8, 2011, Yinghou-1 (which in Chinese means 'luminous fire' or 'shimmering planet') as it was headed to Mars, launched from the Baikonur Cosmodrome in Kazakhstan, riding piggy back on the Russian Phobos-Grunt mission, failed to clear the Earth's orbit. Just days later Chinese state media reported that the China National Space Administration had declared the mission as lost. If this loss of the Chinese mission in a way spurred India to begin its journey to Mars, the loss of the Russian component on the Phobos-Grunt also had its unintended impact by delaying India's second mission to the Moon, Chandrayaan-2 which was being planned in collaboration with the Russians. India has now decided to attempt its next Moon orbiter, lander and rover all on its own, likely to be launched in a few years using the GSLV.

Since 'failure' is a word that does not go down very well with ISRO, the organisation quickly converted these ill-starred events into an immense opportunity and decided to put forward the idea of a first-of-its-kind mission to Mars to the Indian government. In a style that had been perfected over decades, a decision quickly started to take shape, and in just about eight months the formal announcement came through on Independence Day. This was a bold decision by then PM Singh. Not so long ago, decision-making for India's maiden mission to the Moon, Chandrayaan-1, took almost five years. So, while it is indeed true that every baby

From the Red Fort to the red planet. Former Prime Minister Manmohan Singh announcing India's maiden voyage to Mars on August 15, 2012.

Credit: Press Information Bureau, Government of India

step counts in the unforgiving space environments, so does national pride, and what could be more tantalising than the thought that India was presented with an irresistible chance to beat China in reaching the red planet. Banal as the argument may sound, given how Indo-China rivalry has been such a hot topic globally, it is not difficult to imagine the gratification of countless Indians if New Delhi was able to beat Beijing in breaching the inter-planetary frontier. The emotion is even stronger knowing how, to many in the sub-continent, China has been more than just a regional rival, perceived more like the big Asian bully. Some, therefore, suggest that this is the start of a twenty-first century Asian space race where India and China, the two regional rivals, are locked in a modern day inter-planetary marathon. Japan, the third aspirant to reach Mars, is also jogging alongside, whose 1998 maiden effort using a satellite called Nozomi, also failed.

Scientists of course insist that it is only technical reasons that the mission was fast-tracked. Radhakrishnan has said often, and with emphasis, that 'we are not racing with anybody and the Indian Mars mission has its own relevance', asserting that the failure of the Chinese mission and India going to Mars was a mere coincidence. But more on that later, meanwhile on August 3, 2012, a sultry monsoon day, it was the afternoon hours when the Indian Cabinet, chaired by then Prime Minister Singh, had a historic meeting in the South Block offices on

India's Mars Mission is riding on the success of the earlier mission to the Moon in 2008. Scientists seen jubilant on the successful launch of Chandrayaan-1 on October 22, 2008 in Sriharikota.
Credit: Pallava Bagla

Former Prime Minister Manmohan Singh inaugurating the Indian Institute of Space Technology, Thiruvananthapuram, along with the two stalwarts of ISRO, G Madhavan Nair (L) and K Radhakrishnan.

Credit: Press Information Bureau, Government of India

Raisina Hill in New Delhi, where it unanimously approved the Mars Orbiter Mission. As is customary, usually a Cabinet Minister briefs the media about decisions taken at every such meeting, and there was palpable anticipation that the formal announcement of the decision to go to Mars would be made. Union Minister Mr Kapil Sibal came out and conducted a briefing about key decisions taken at the high-level meeting, but it was an anti-climax for science correspondents, for he did not have much to say about Mars. Obviously, some hard questioning was warranted, and on being quizzed by the author (Pallava) Minister Sibal said, 'I am not authorised to speak about Mars.' One thinks the truth lay elsewhere and his tight-lipped response was perhaps more to do with a need to keep the news under wraps, saving the drama for the Prime Minister's Independence Day speech. In the next few days, there were obviously a lot of efforts to ensure that the news of the decision didn't leak out and that the big news of India's inter-planetary leap would be the highlight of the 2012 Independence

Day declarations. All said, news did leak out that the mission to Mars had been given the green signal. The MOM Executive Summary Document,[1] prepared as late as July 2012, set out the plan: 'The Mars Orbiter Mission (MOM) is ISRO's first inter-planetary mission to planet Mars with an orbiter craft designed to orbit Mars in an elliptical orbit. MOM can be termed more of a technological mission than a science mission considering the critical mission operations and stringent requirements on propulsion and other bus systems of the spacecraft. MOM has been confirmed to carry out observations of the physical features of Mars and carry out a limited study of Martian atmosphere with five payloads finalised by ADCOS (Advisory Committee on Science).' But as the ADCOS Chairman, UR Rao, legendary former chief of ISRO caustically put it, 'small instruments give small science'; he also did not hide his views when he suggested his preference was to go visiting Mercury, a virgin planet, almost unexplored.

It is evident that the reasons to go to Mars are complex. Both geo-political and planetary configurations have played on the minds of Indian policy makers who took the Martian plunge.

© ESA/DLR/FU Berlin (G. Neukum)

The Indian mission to Mars was cleared within weeks of the maiden Chinese mission Yinghou-1 riding piggy back on the Russian Phobos Grunt having failed. Seen in the photo are the proposed landing sites of the Phobos Grunt mission on one of Mars' Moons called Phobos.

Credit: European Space Agency

S K Shivakumar, Director of ISAC in Bangalore admits that the design and development of the Mars satellite was indeed 'fast tracked.' Of course it is scientific and technical reasons that the space community is most comfortable talking about, perhaps because of the feel-good stuff it obviously is. After all, Mars is a popular planet. The largest volcanoes and deepest canyons of the solar system are there, so are deep channels and rocky outcrops and impact craters, sand dunes and Martian ice and clouds. And it is said that men are from Mars, while women are from Venus! So, its associations are all male, the red planet, with its rocks and soil and sky having a red hue. It orbits between the asteroid belt and the Earth. Since there are characteristic criss-cross lines as patterns across the surface of the planet, there has been constant speculation that these lines are actually irrigation channels, built by those who live there, and there is also the observation of apparent seasonal colour changes, which seem like there could be vegetation; so the 'life on Mars' debate is one of the hottest in the space exploration world. Though, as yet life on Mars has only been found in cartoon films. As the Executive Summary report put it, 'Beyond the Earth's vicinity, Mars is a natural target of study in India's solar system exploration programme. Of all the planets in the solar system, Mars has sparked the greatest human interest. Its orbit lies between the asteroid belt and Earth. For ages, humans have been speculating about life on Mars. The conditions on Mars are believed to be hospitable since the planet is similar to Earth in many ways. Like Earth, it has an atmosphere, water, ice and geology that all interact with each other to produce the dynamic Martian environment. Mars has surface features reminiscent of both the impact craters of the Moon and volcanoes, deserts and the polar ice of the Earth. But, the question that is to be still answered is whether Mars has a biosphere or ever had an environment in which life could have evolved and been sustained.'

Of contemporary space explorations globally, missions to Mars are considered of prime importance and therefore planned and implemented most often. Unravelling mysteries, analysing Martian meteorites and remote sensing data from past missions to the planet are what create the knowledge base about it. It is surely exciting to think about past evolutionary processes. And the knowledge leads to more questions, the affinities between Mars and Earth also because they are, after all, next to each other in the solar system; and there are other affinities, like the weather is apparently quite like that on Earth. There is air, there is water, there is ice. Like the other three rocky or terrestrial planets of the solar system (Mercury, Venus and Earth), Mars has an interesting geology and is half the radius of the Earth, it also has two small, irregular Moons – Deimos which means panic in Greek, and Phobos, which means fear. But to critics, this is a sub-optimal mission with very limited scientific objectives, and they were not silenced by the economic argument of this being the most cost-effective of all missions of half a century of Mars exploration. Interestingly, when the Moon mission was

being planned, Chandrayaan was announced from the Red Fort by Mr Atal Bihari Vajpayee who was Prime Minister in 2003, and the build-up was the same. And Prime Minister Singh, as it seems, took the cue, delivering his speech which was actually written in Urdu! But who could match the politician visionary in Vajpayee, who made sure to call the Moon mission Chandrayaan-1, whereas, the bureaucrat in Singh merely called it Mangalyaan, not thinking at all about any sequels!

What's in a name you might think? When the first budgetary clearances of ₹ 1,250 million were given for the Mars mission on March 16, 2012, at the time of the 2012–13 budget presented by the then Finance Minister, Pranab Mukherjee, it was officially recorded: 'Mars Orbiter Mission envisages launching an orbiter around Mars using the PSLV XL during the November 2013 launch opportunity. Mars Orbiter will be placed in an orbit of 500 × 80,000 kilometre around Mars and will have a provision for carrying nearly 25 kgs of scientific payload.' In what might be some 10 kgs of documents running into thousands of pages embedded in a book with orange covers called the *Expenditure Budget* laced with tables and millions of numbers, was a small annotation that announced this remarkable plan giving financial sanction. 'What a mangal announcement', was the observation on March 16, 2012, when one of the authors was in conversation with jubilant officials in the Prime Minister's office thinking about what name the mission would possibly get. How India's maiden mission to the Moon was subsequently named Chandrayaan, and how this might perhaps become Mangalyaan was alluded to, and of course some months later, in his speech from the Red Fort, the PM irrefutably referred to the mission as Mangalyaan. This was a deviation from his written speech, and that is why many people who went and revisited the written speech questioned whether it was ever christened this way by Prime Minister Singh. There is also a small technicality that made ISRO formally stick to calling India's maiden mission to the red planet as the 'Mars Orbiter Mission (MOM)'. It seems DOS officials had sent a list of names to the Minister of Space, a portfolio always held by the Prime Minister of India, to tick the name of his choice. As it happened, Prime Minister Singh apparently never returned that piece of paper articulating his preference, perhaps owing to his many other scam-based preoccupations, and this led decision-makers within the space community to safely stick to referring to it as MOM. This contrasts quite sharply with how the Moon mission was named, when in 2003 a list of names was sent to Prime Minister Vajpayee for naming the Moon mission, ISRO's preferred name was 'Somayaan', but astute officials around Vajpayee renamed it as 'Chadrayaan' and the author (Pallava) was told by Mr Sudheendra Kulkarni, then press advisor to the Prime Minister, that Vajpayee himself added, in his own hand writing, the suffix '1' after Chandrayaan, while wondering how a big country like India could make only one visit to the Moon. On the issue of naming the Mars mission, Radhakrishnan has explained that a choice was made to 'wait for an opportune milestone in

The first official public announcement by ISRO about the Mars Orbiter Mission on September 11, 2013. At the press conference (L to R): M Annadurai, Programme Director; SK Shivakumar, Director, ISRO Satellite Centre; and S Arunan, Project Director.

Credit: Pallava Bagla

the mission to invite entries from schools and colleges to find a suitable name, but this is yet to happen.' This led decision-makers to stick with the informal names, MOM and Mangalyaan.

The geo-political hues and undertones of space programmes are well-known and affect national priorities. Through whispers in the corridors of power in the government, it is known that after India's Moon mission was announced on May 11, 1999 and finally executed only in 2008, China had ample opportunity to overtake India. This was, for obvious reasons, a disturbing fact for the Indian government and one that could not be forgotten easily. It is a well-kept secret that the government has remained cross with ISRO on Chandrayaan-1 because we were beaten by China in the Asian space race to the Moon, and it may sound childish, but possibly as a mark of showing his displeasure, Prime Minister Singh never ever gave formal audience to the Chandrayaan team. The absence of this ceremonial, but morale-boosting protocol usually extended by the Prime Minister for much less significant satellite launches is clearly a sign for the institution that it was being ticked off for this expensive delay, that some suggest dented India's national pride! A costly mistake ISRO did not want to repeat.

So does India's Mars mission mark the beginning of a new Asian space race? India sees the Mars mission as an opportunity to beat its regional rival China in reaching the planet, especially after the 2011 failure the latter suffered. China has beaten India in space in almost every aspect so far; it has rockets that can lift four times more weight than India's, and in 2003, successfully launched its first human space flight which India has not yet embarked on. China launched its maiden mission to the Moon in 2007, ahead of India. So if India's mission succeeds, it will have something to feel proud about. Since 1960, 51 missions to Mars are known to have been launched, more than half of which, 27 to be precise, have failed; the journey is risky and Mars has traditionally been an unfriendly environment for humans! So much so, that there was once a colourful belief that a 'galactic ghoul' or monster ate up all missions heading towards the red planet. And no nation, apart from the Mars Express mission, Europe's maiden venture to Mars representing a consortium of 20 countries through ESA, has succeeded in a maiden venture.

Maybe these histories are what zipped the lips of space officials for the longest time even after Prime Minister Singh made his announcement about India going to Mars. There are

In happier times: G Madhavan Nair and K Radhakrishnan.
Credit: ISRO

other incidents that were creating a somewhat 'where-angels-fear-to-tread' kind of atmosphere for the space community. Not long before, ISRO chief Radhakrishnan and his predecessors, and the space community in general had been affected badly by the S-Band scandal. Individual reputations and that of the institution were perceived as being at risk, having taken a serious blow after decades of a squeaky-clean track record. So, the need for a feel-good piece of work was fairly urgent. But in its wake came further controversy. The announcement gave an opening for G. Madhavan Nair, former ISRO chief, often referred to as the 'Moon man', who had by then become a serious detractor of Radhakrishnan, often referred to as the 'Mars man' to extract his pound of flesh after being singed in the Antrix-Devas mess in which he had been black-listed along with a few others. Mangalyaan would be 'a national waste', Nair said, describing the Mars project as 'a half-baked, half-cooked mission being attempted in undue haste with misplaced objectives.' It doesn't take much intelligence to note that Nair's name is often left out of high-powered conversations in ISRO, despite it being an institution known to celebrate all its past leaders.

At the multiple institutions of ISRO across the country, floor action begins far ahead of final and formal approvals. This is to be expected, given the kind of lead time complicated space engineering projects require. So, the seed money is used to keep the floors warm, and this ensures that once final approvals come in, no time is wasted in tinkering with the elementary stuff of any project. Since this is their way of working, it is sometimes difficult to figure out exactly when a particular project actually began, and so, the dash of the Martian marathon should also be viewed in that framework. Previous heads of the institution had mentioned ISRO's intent to go to Mars, and Radhakrishnan mooted the idea in 2010, launching a serious effort by commissioning feasibility studies, initiating dialogue among the space community and leadership and consolidating efforts as a run-up to the formal interfacing with the government. To those outside ISRO it remained an internal ISRO dream since the government was still not willing to commit tax payers' money for this mission.

The dream picked up steam with the loss of the Chinese mission in November 2011, and as described earlier, this was like a door getting opened for ISRO to sneak in a proposal

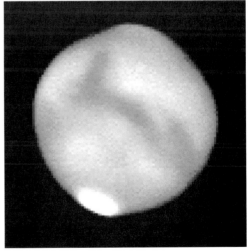

for its Mars ambition. All those who dabble in the strategies of governance, decision-making and financial planning have a practiced knack for pushing proposals when they believe the time is right, and ISRO is hardly an exception. A cold, foggy day dawned on December 21, 2011, and what is usually quite a secretive and closed-door, a meeting of the high powered Space Commission was held in the New Delhi office of the ISRO Chairman. What an unlikely location it is, the New Delhi office of ISRO, contrasting starkly with the organisation's high profile and remarkable offices and buildings in Bangalore. In modish Khan Market in central Delhi, a locality where the city's well-heeled shop, is a rather shabby, non-descript government building called Lok Nayak Bhawan. Best known for its lamps and lighting

The red planet as seen by India's highest astronomical observatory at Hanle in Ladakh. This image of Mars was captured on August 27, 2003.

Credit: Indian Institute of Astrophysics

Straight out of sci-fi, the high tech Mission Operations Centre in the heart of a nondescript industrial neighbourhood of Bangalore.

Credit: Pallava Bagla

market housed on the ground floor, this building has the ISRO office on the third floor that one reaches using a rickety elevator that could do with a clean-up job. Tea is served in the spartan 'Delhi camp' office of the Secretary of the DOS in glistening bone china cups, with salted cashew nuts for accompaniment. This is where some of the most powerful bureaucrats and scientists deliberated on a proposal brought in front of them, detailing India's plans to head to the red planet. There can be no doubt that the failed Chinese mission had its overhang on the meeting. By the time Radhakrishnan, who chaired this meeting, emerged, India's mission to Mars had been given the green light. The dilemma in front of the ISRO chief known for his 'managerial skills' was how to complete the mission in a short span, considering that on that particularly cold day, financial allocation was still missing. It finally came only in warmer March 2012. But like on many other occasions, ISRO's legendary forward planning came to its rescue and the organisation decided to take the challenge head-on and made all efforts to complete the mission on time and without cost over-runs. Unbelievable and almost unthinkable in the red tape-laden government file pushing that India is famous for, but then ISRO is different! That it was a mission worth experimenting was only one factor; that ISRO also needed a morale booster after what it had gone through in the last several months because of the S-Band scandal was clearly another. Mangalyaan offered the right platform, according to Mr Shivshankar Menon, National Security Advisor in the erstwhile Manmohan Singh-led Cabinet and Member, Space Commission. Early considerations veered towards planning a fly-by mission, and this clearly has advantages when a first-time inter-planetary mission is to be accomplished. But, according to a feasibility study report on Mars prepared by Dr V Adimurthy, who is a Satish Dhawan Professor at ISRO, and a panel of experts, and readied in July 2011, made available to the authors, there was greater benefit in mounting an Orbiter-based mission as it would be more advanced in terms of complexity and 'will have wider scope(s)'. This would 'provide the opportunity to study Mars for a considerably long period as compared to a spacecraft in a fly-by mission.' This led the scientists to recommend an orbiter mission vis-à-vis a fly-by.

So an orbiter it would be. Overnight, what ought to have been a marathon to Mars, became a 100-metre sprint because the powers that be had determined the launch date, egged on by clear geopolitical earthly reasons, and this meant a small, but critical launch window in the month of October-November 2013, had to be met come what may. It so happens that once every 26 months comes a window of opportunity to relatively easily reach Mars in terms of celestial orbital alignments, according to the scientists. This placed an incredible time constraint and the mission had to be readied in 15 months. This meant that a lot of Indian innovation or *jugaad* had to be put to creative use in getting the mission up and running. For example, the payloads had to utilise products procured from the market and re-engineered for making them fit for their purpose. The report also clarified that the PSLV-XL could be the vehicle of choice. As Radhakrishnan says, 'PSLV-XL could lift an orbiter spacecraft towards its journey to Mars at the next opportune time in November 2013. A stupendous task but a possible turning point for India.' The mission to Mars presented several challenges to the multi-disciplinary teams at ISRO – responding to the need to remain as indigenous as possible, ensuring rocketry for such a long journey was in place and robust enough, creating scientific instruments that could remain light and utilitarian, and building the satellite itself. And another challenge that India is very familiar with – to do all of this within a tight budget; this mission was highly cost-effectively planned at ₹ 4,500 million or about USD 70 million, and with 500 engineers and scientists on the job, was meant to test several technological aspects of a deep-space inter-planetary mission. From orbit manoeuvres to navigation in deep space and such incredibly long-distance communication; and of course the scientific objectives that would be fulfilled by the payloads on board the MOM satellite.

While there is no doubt that the ultimate goal of this mission for India is to get to Mars, but the journey there – long and quite arduous – is equally challenging and one that is being watched, monitored and controlled each and every minute by hundreds of technicians and space engineers and scientists. In short, even the journey itself would be a great achievement, and that is why key scientists on the mission call MOM a 'technology demonstrator'.

This meant that there was no time really for all the wonderful international institutional partnerships that make everybody feel good. Partnerships and collaborations would take several years and India didn't have the luxury of time. So it would be a lone journey. Going alone is an adventure, but a difficult one. To Annadurai, this became a clarion call, and he anyway saw the Mars mission plan as a natural extension of Chandrayaan-1, that after going to the Moon, you try to reach a planet. The 100-metre dash would also mean an incredible amount of *jugaad*, or local ingenuity and frugal engineering put to good use. This *jugaad* challenge meant a whole lot of things. Most importantly, the GSLV was unavailable so the only option was the old favourite, PSLV, and that was a sub-optimal decision, because it can give a thrust that is enough only to go around the Earth, so what they had to do was to use the on-board fuel of the satellite to get the extra kick required by firing it in stages. This was a big compromise since most global missions save precious satellite fuel that can be used to maintain the Martian orbit of the satellite and to extend the life of the mission once it reaches its destination. Anyway, it seemed, India's smaller rocket was the only option available in 2013 since the GSLV was a still the 'naughty boy' of the Indian rocketry programme, finally being tamed only in 2014. Radhakrishnan explains that the 'PSLV can inject MOM into

the orbit as required for minimum energy transfer from Earth to Mars. GSLV requires a restartable cryogenic upper stage to do this, which we did not have.' To the soft-voiced Dr P Kunhikrishnan, PSLV Project Director for the Mars mission and man of 11 missions, it was a very special time because the PSLV completed 25 missions, and he felt particularly proud that the twenty-fifth mission happened to be the Mars mission. To this man, each mission is unique and every mission uses the versatility of the PSLV in a different way, although it was initially developed for launching satellites into a low Earth polar orbit. With its totally indigenous Vikram computer, the PSLV once carried as many as 10 satellites. On April 28, 2008, the Indian space agency created a unique world record by successfully launching 10 satellites in one go, shattering the previous record of a Russian rocket which was actually a modified intercontinental ballistic missile that launched eight satellites in one shot on April 17, 2007. This was really not an easy record to conquer, since launching 10 satellites into outer space has its own challenges, including the chance that collisions may happen. As one ISRO official told the author (Pallava), it was almost like when a school bus drops children home after school, one after the other. When such a tricky operation starts, the rocket is already travelling at a speed of 7.5 kilometres per second or blazing at 27,000 kilometres per hour and everything has to happen in a flash, which is why immense precision is required. Meanwhile, the *jugaad* continued with the satellite bus being prepared by Hindustan Aeronautics Limited (HAL), Bangalore through its Aerospace Division, and the MOM satellite structure which was an assembly of composite and metallic honeycomb sandwich panels with a central composite cylinder was then filled up at the satellite centre at ISRO. This complex piece of electronics broke all previous records in terms of the time taken to make it, and cost ₹ 1,500 million, forcing many jaws to drop. How could anyone even attempt making a satellite heading to Mars at under USD 40 million since in contrast NASA's Curiosity cost a whopping USD 2.5 billion.

In the Adimurthy report, the scientific objectives of the mission were spelt out – 'to investigate the physics, chemistry and dynamics of the upper atmosphere and the escape of volatiles into space, and to investigate the chemical, mineralogical and geo-morphological features of the Mars surface.' For this purpose, five payloads, each one completely indigenous, were pulled together, again using innovative, fast-track methods. The Mars Colour Camera (MCC) is the 'eye of the mission', meant to capture images and information about the surface features and composition of the Martian surface, besides being useful for monitoring weather. The Thermal Infrared Imaging Spectrometer (TIS) will map resources and mineralogy. The Methane Sensor for Mars (MSM) will measure methane, that ultimate provocateur gas that entices us to think of the possibility of life, and the Lyman Alpha Photometer (LAP) will check for the Deuterium and Hydrogen ratio, helping scientists understand the loss of water from the planet, and leading to answers as to why there is so much water on Earth. The Mars Exospheric Neutral Composition Analyser (MENCA) will taste and smell the particulate environment. Each of these instruments has a specific purpose for being on the satellite. Sampling the thin Martian atmosphere, about which we know so little, is critical, but the global scientific community is very excited about India's efforts to send the first dedicated methane gas sensor to Mars. The presence of methane gas, also called 'marsh gas,' on Earth is one of the clinching signs of the presence of carbon-based life forms. So in a way, without even landing on Mars, India hopes to provide an answer to that big question – 'Are we alone in this Universe?' The need to measure methane is of course there, as is the need to image the surface, find out what it is like, and measure and analyse the chemical composition of the Martian environment. Selecting each of these was a difficult task and one that was possible because of ISRO's culture of rigorous technical reviews through diverse committees. In different parts of India, in different institutions, thousands of men and women (many more men than women, unfortunately) are in some way or the other, linked with the Mars Orbiter Mission. It must be said though, that

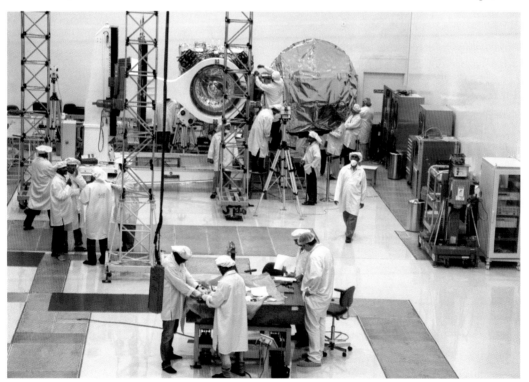

The clean room of the ISRO Satellite Centre where Mangalyaan was born in 15 months.
Credit: Pallava Bagla

Ready for lift-off. K Radhakrishnan, Chairman ISRO (Centre), seen with the PSLV in the background in Sriharikota.

Credit: Pallava Bagla

the person who does final checks on the satellite is a woman, so is the young engineer who navigates the massive 15-storey PSLV from its home in the vehicle assembly building to the launch pad at Sriharikota.

As the mission came together, time was of the essence. Frugality was an equal driving force, not just in terms of financial resources available, but also the amount of weight that could be carried on board (just 25 kgs in all). Finally the five payloads were weighing less than 15 kgs in total and gave ISRO the opportunity to give more space for Indian scientific instruments or take an international partner. In the summer of 2012, the author (Pallava) confirmed with scientists at the Smithsonian Institution, Washington DC, USA who work on NASA's Mars Curiosity programme, who not only expressed surprise that India was undertaking a mission at such short notice (the Indian mission had not been publicised much globally at the time), but also felt that if there was an opportunity they would have come forward to collaborate, that the American Martian explorers were more than willing to partner on the scientific mission. But then, India had no time, with Radhakrishnan already visualising himself as the lithe and sharp Arjuna of Kathakali, his favourite pastime, keeping his eyes focused only on the reflection of the fish in the water, aiming his twenty-first century bow to take a shot at Mars. And the luxury of being off the mark did not exist. Having said that, General Charles F Bolden, Administrator of NASA and former astronaut, said to the author (Pallava) in a long-ranging and in-depth interview that the Indian Mars mission was indeed supported by NASA's deep space tracking systems. A paid service India accessed from the Americans that cost ISRO about ₹700 million or about USD 11 million. This service was needed since there would be critical times when Mangalyaan would not be visible to the Indian deep space network located near Bangalore in the village of Byalalu, where a giant 32 metre dish antenna tracks the MOM day and night.

Meanwhile, in planning for the long-distance, inter-planetary journey, on-board autonomy is critical for a space craft – the ground systems are way too far off for timely and efficient communication. The communication system had to be altered and adapted to factor these vast and unimaginable distances, and to ensure effective navigation so that the craft would correct itself on a journey that is a thousand times longer than what Chandrayaan-1 had to undertake. On board autonomy in communication would be critical, as also enhanced power generation capacity; sending 20 hours of data would take 37 hours, and for every hour of data that the MOM would record, it would take two hours to send it back to ground stations.

What all this meant was that dedicated teams were on the job and people worked 24 hours a day. Teams within institutions, and teams across institutions – people working on the satellite itself in Bangalore, three of the payloads at Space Applications Center in Ahmedabad, one payload at Bangalore, one at VSSC in Thiruvananthapuram, and the rocket at VSSC. Besides, scientists at the Physical Research Laboratory (PRL) in Ahmedabad focused on what possible scientific discoveries might emerge from the mission, and what could be done through these discoveries. It should also be noted that Mangalyaan remains a mission with no one designated 'chief scientist' for the mission. For the Chandrayaan-1 mission, planetary scientist Prof JN Goswami, Director of PRL was the 'chief scientist' and it really paid dividends when his name as a co-author was part of the landmark scientific papers that announced the presence of water on the Moon which was till then considered a parched environment. It was published on September 25, 2009 as a cover story in the prestigious American weekly, *Science* magazine published from Washington DC.

So, is the Mars mission a hasty experiment to prove a national point? More so, since anchors on Indian television channels screamed in endless debates on what should be India's priority – Rockets versus 'Rotis'; Mars versus Mal-nutrition! Or even a scientific risk, knowing orbital positioning issues. Or, as Jean Drèze, a development economist at Allahabad University and the Delhi School of Economics, University of Delhi put it as 'a part of the Indian elite's delusional quest for superpower status?' Ambassador Shivshankar Menon, former National Security Advisor, has little patience for the self-doubt and the flagellation over whether India

should do this or that, and how the country can afford space missions against its abysmal human development indicators. 'We have heard these arguments since the 1960s, about India being a poor country not needing or affording a space programme. If we can't dare to dream big it would leave us as "hewers of wood and drawers of water!" India is today too big to be just living on the fringes of high technology.' There's something to be said about the fact that big dreams are for everybody. Indeed they are; as they say, 'no dream is too big!'.

NOTE

1. ISRO Mars Orbiter Mission, Executive Summary, Department of Space, Bangalore. Government of India, July 2012.

Chapter 3

Unravelling Martian Mysteries: The Known and the Search for the Unknown

The chances of anything man-like on Mars are a million to one.

– H.G. Wells, *The War of the Worlds*

Exploring Mars is a bit like doing brain surgery through a mile-long soda straw. At an average distance of fifty million miles from Earth, with a one-way radio message time of twelve to twenty minutes, roving the dry, treacherous surface requires the utmost planning and careful execution. One false move can end a mission in seconds, and there are rarely many options to correct a mistake. That is why people who dare seek the truth about Mars are so remarkable.

– Rod Pyle, *Destination Mars: New Explorations of the Red Planet*, 2012

The year is 1898. Herbert George Wells, the Britisher who later becomes famous as the father of science fiction, writes the iconic *War of the Worlds*. This is the first-ever story that alludes to a Mars that has some 'people' living there – Martians. The book spoke of Martians invading Earth, and opened the floodgates in many ways. The planet Mars and its mysteries became the most popular themes of extra-terrestrial (ET) fiction, and the 'anybody out there' idea got rapidly cemented, with fair contributions from a series of popular films made from the 1950s onwards. It remains a catchy, saleable theme in today's world. But much before the movies, George Orson Welles, American actor, director, writer and producer conceptualised and aired a radio drama based on the book and this was a rage. Of course, it was only 1938 and a wave of panic spread across America, with ordinary people worrying about an alien invasion of their beloved Earth. This notion of another community of living beings either similar to or very different from earthly humans, somewhere else in the Universe, held a powerful sway on human minds the world over, and it still has, into the twenty-first century. *War of the Worlds* was made into a much-watched movie in 1953. Another early milestone in such story-telling was a short story collection called the *Martian Chronicles* published by Ray Bradbury in 1950, with gripping narratives of humans escaping from a devastated Earth and reaching Mars to colonise it, only to clash with 'aboriginal Martians'. Children of several generations spread across the world have spent years in the belief that aliens, ETs and Martians can land anytime, and this has spawned excitement in some, fear in others, and scorn in many more. After all, human imagination is indeed a great force, seting us apart from other animals. But then there is another trait that makes a difference, and that is the human ability to focus on science and scientific enquiry, and the driving desire and skill that the human brain has to relentlessly question, study, research and compile information till giant bodies of work get built and keep growing. It is this nature of earthly beings that has created and nurtured knowledge societies; and interestingly, this thread has run along amiably with the other thread, that of our dramatic imagination and creative thinking. Somewhere along those journeys, the two meet and the intricate fabric of modern society gets woven with the coexistence of science and society, rationalism and religion, fact and myth. In the vivid expression of General Bolden, 'Every single flight, I was looking for ET! You're always hopeful that you will be able to contribute to answering the question, are we alone, which is the age-old question for humanity. I would love to be the first to find some evidence that there is life elsewhere in our Universe, but as yet there is no evidence of ET though.' Speaking to the author at length in an exclusive interview,[1] this space exploration leader of the US space programme has perhaps captured the essence of the dilemma that is a constant in human society. Anyway, the second part of the twentieth century was distinctive by its incessant pursuit of planetary and space science, leading to an enormous body of knowledge that soon began to speak of Mars as an airless, bone dry, and somewhat dead planet that most likely could not support life, elbowing the Martians out in one fell swoop.

That is what some would like to believe. But it isn't so simple, and while mission after mission has brought back pictures, data, interpretations and counter-interpretations, the stories have also endured, allowing people to continue their lingering romance with the red planet and its

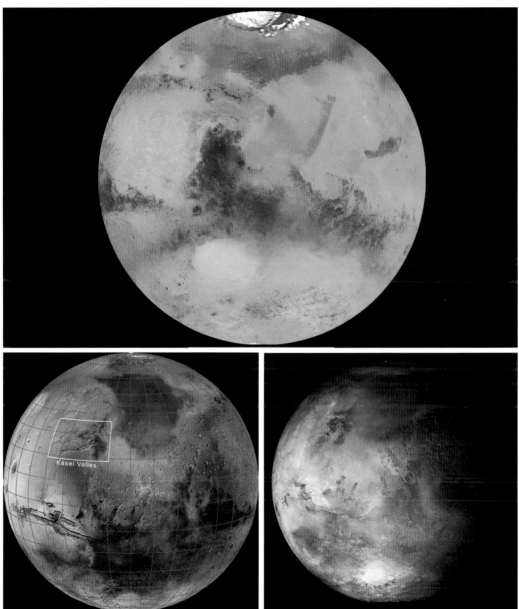

Kasei Vallis

The red planet and its many mysteries. The search is on to better understand Earth's neighbour as the world prepares for sending humans to Mars in the next few decades with the distant ambition of colonisation.

Credit: NASA and ESA

NASA's Curiosity Mars Rover as seen on the Mars surface in an artist's impression.
Credit: NASA

endless mysterious possibilities. This has kept warm the search for any signs of life on Mars, further fuelled by the similarities between Mars and Earth. There are polar ice caps and clouds in the atmosphere, seasonal weather patterns, volcanoes, canyons, and other recognisable features. There seem to be physical features that resemble gorges and riverbeds, shorelines and islands, and the possibility that great rivers once meandered through the Martian landscape. Of course, there are differences too, like there are no oceans on Mars, and the entire planet is probably dry. Overall, conditions are most likely very extreme. There is no Ozone layer or a magnetic field to protect the planet, and charged cosmic particles, as well as ultra violet rays enter deep into the Mars atmosphere. All these differences could be partly because the precursor materials that initiated planetary development of both the Earth and Mars could be different, and partly because their evolutionary paths are different. Meanwhile, Mars is a much smaller planet, with its diameter being about half that of the Earth, but it takes about the same time to go around the Sun as does the Earth. The force of gravity is much lower on Mars, being one-fifth of that on Earth. A planetary neighbour, friendly or otherwise, Mars is one of the few members of the Solar System that is visible from Earth just by the naked eye, despite being the second smallest planet of the Solar system, after Mercury. Its colour is also distinctive. The Moon is by far the brightest, Venus a close second, and these two are also visible to the naked eye. So, gazing at Mars, people can let their imagination fly, just because they can see the planet and the similarities that keep getting mapped. It is like if there is *terra firma*, then can life be far behind?

All this makes for ideal ground for one of the hottest topics of present day space exploration, the colonisation of Mars. Like all instances of colonisation, here too there seems to be an element of aggression, perhaps because of historical linkages. Over millennia in the history of the Universe, Mars has stood variously for war, terror, fear, fire, the colour red, murder, destruction, a burning coal, grief, and bane, evidently all synonyms of some negative emotion or the other. These are but some of the words that come up in etymological deconstructions of the word Mars. In almost time-honoured tradition, the planet has also been worshipped by almost every culture in the world. In Greek mythology, Mars is the son of Zeus and Hera, and is called Ares. Homer wrote in the *Iliad* that Ares was bloodthirsty and full of anger and wrath, yet a coward to boot. This is pointed out as the reason for Zeus and Hera not being very fond of their son. Ares had three children, two of whom were called Phobos (which means 'fear') and Deimos (meaning 'terror'). Today, two Moons of Mars are named after Ares' children. It is the Romans who first named the rather 'bad' planet as Mars, and according to them Mars had two sons, Romulus and Remus. Naming their ancient, now iconic city Rome, after the former, they perhaps inherited their ancient warring ways through this connection with Mars. There are other connections as well, like the city of Cairo being named after Mars;

Al Qahira was an ancient Arabic name for the planet, and this became Cairo by modern interpretation. Mars also was called *Angarakan* in Sanskrit.

Rough and rogue-like, Mars seems to exercise negative influences; this is reflected even in current day India where Hindu astrological practice requires that every person has his or her own astrological chart mapped according to the nature of planetary constellation at the time of birth. So, a person born during the time when Mars is in a certain position is called a *Manglik*. The superstition is so strong that marrying a *Manglik* is considered highly unfavourable by those who follow such practices, believing that such a marriage can only be a den of troubles, with tensions and discord leading to increasingly serious problems as the marriage progresses. In an effort to find a solution to this problem, a ritual has been created where a *Manglik* is first married off to a banana tree, or a *peepal* tree, or a silver or gold idol of a Hindu god or goddess. After this ceremony, the actual human marriage is solemnised. The mythological descriptions in Hinduism that relate to the birth of the deity *Mangala,* which stands for anything 'auspicious' in Hindi and Sanskrit, are equally aggressive. Born out of the sweat of Shiva, the Hindu god of destruction, *Mangala* is described as having four arms, each arm carrying a trident (or a *trishul*), a mace, a lotus and a spear. He is shown sitting on a ram.[2] How *Mangala* came into existence is also a mythological story, that of the demon *Andhakasura* who received a boon of immortality. The boon was that if he got injured during war, another *Andhakasura* would rise from each drop of his blood that fell on Earth. As a result, the demon was nearly undefeatable in war. Driven by this feeling of immortality and unparalleled strength, *Andhakasura* became a trouble-maker for the Gods, often harassing them, and the holy sages. Deciding to do something, the gods sought an audience with Lord Shiva, beseeching him to put an end to their suffering by killing the demon. A great fight ensued in ancient *Avanti* which is present day Ujjain in Madhya Pradesh and Shiva, known as the Hindu god of destruction, locked *Andhakasura* in a tough, one-to-one battle. It was tough because *Andhakasura* had that special boon. But in the final round, Shiva wiped his brow of sweat, with beads falling on the battleground. This sweat gave birth to *Mangala*, who drank the blood that Shiva drew from the demon's body after piercing his heart with his trishul. Since all the blood was gulped down by *Mangala*, not a drop fell and there were no more *Andhakasuras* to be born. But *Mangala* was born, red and aggressive. A reputation that abides till today.

Maybe that is why Mars has been an obsession with astronomers over the centuries. Armed with the earliest and most rudimentary of telescopes, early star gazers were very active, but even then it is hard to say who actually first spotted Mars, but right from pre-historic times, starting with the Egyptians who were known for their astronomical skills, and observed Mars back in the second millennium BC, the planet was certainly being observed. The Chinese have records

America, Russia and the European Space Agency (ESA) have successfully conducted missions to Mars. Seen here is an artist's impression of the Beagle spacecraft sent out by the United Kingdom as part of ESA's Mars Express Mission in 2003. This spacecraft crash-landed onto the surface of Mars.

Credit: ESA

of the movements of Mars even before the founding of the longest-running Zhou Dynasty in ancient China, and so did the Babylonians who worked out mathematical ways to predict planetary positions in the future. Around 356 BC, Aristotle observed that Mars was moving behind the Moon. And then, in 1610, the telescope was invented by Galileo Galilei who made his first observations, saying that Mars was not completely round. Over these infinitely long periods of time, the study of Mars became more and more important, and what started back then is till date a serious activity. The following table provides milestones in the history of science and scientific inquiry and it is indeed strange that even so many hundreds of years ago, astronomers were not saying anything very different! Considering the enormous amount of financial and other resources that are now being spent by nations just to study Mars, the returns ought to be such that they boost our knowledge by leaps and bounds. No doubt that some early astronomers did let their imagination run wild, with their visual observations straying far from the facts to simply fit some pictures that had been conjured up in peoples'

Table: Unravelling the Martian Mystery: Early Sightings

1610	The telescope is invented by Galileo Galili, and Francesco Fontana, an Italian astronomer, prepares the first drawings of the Martian surface.
1659	Christiaan Huygens (Holland), better known for developing the wave theory of light, observes and draws Syrtis Major (one of Mars' dark regions that are now called maria) on a sketch of Mars. It was called the Hourglass Sea earlier, and it looked like a V. He also observes a bright and shining spot – the polar ice cap. By observing this ice cap, Huygens projects the period of rotation of Mars, at a nearly accurate 24 hours.
1660–66	Giovanni Domenico Cassini,[3] whose name was later changed to the French Jean-Dominique Cassini, makes sketches, like paintings, with not much detail. But he also calculates the duration of a day on Mars, and measures the planet's period of rotation to be 24 hours and 40 minutes, more accurate than Huygens and missing current day accuracy by three minutes! He finds a star dimming even before Mars could cover it, making him deduce the presence of an atmosphere of a substantial nature on Mars. He also notices how there are white spots on the surface of Mars, which today are called the polar ice caps.
1719	Giacomo Filippo Maraldi, Giovanni Cassini's nephew, conducts some observations of his own. After many years, Maraldi is convinced that the shapes of some maria change over time. He thinks this is evidence of clouds that sometimes obscure the surface. He also jots down observations about changes in the polar caps and speculates that this could be evidence of seasons – ice from the polar caps supposedly melts during the 'summer' and freezes again during 'winter'.
1777-83	William Herschel 'Britisher' notes the axial tilt of Mars and deduces that it should have seasons and an atmosphere not very different from that of Earth. This was a time when the telescope quality had improved greatly. Herschel builds a giant 36-inch reflecting telescope in 1777.
Late 1800s	This is the time that sees major activity with improved optics and larger telescopes, and better tracking mechanisms.
1873	Flammarion questions the human analogies of the red blotches of Mars being land and soil, the green, plants, and the white snow saying that humans seemed to speak of plants, snow, seas and clouds on Mars as though we have seen them.
1877	Asaph Hall creates maps of Mars and becomes the first to note its two Moons and names them Phobos and Deimos
1894–96	Giovanni Schiaparelli makes high quality maps of Mars, using a nine inch refracting telescope in Brera, Italy, and these were the first where the famous Martian canals were depicted. He famously called them 'canali'. But his analysis mentions both possibilities: one that these were the result of natural planetary evolution, but the second more tantalising one of how they were too perfect geometrically to be simply the work of nature.
1894	William Wallace Campbell, American astronomer, conducts spectroscopic analysis and deduces that little, if any, water and oxygen are present.
1894	Percival Lowell builds an observatory in Arizona, naming the place Mars Hill, writes prolifically, and works on the canals' concept by saying that the lines must be artificial canals because they are straight, uniformly wide and go from one point to another. He explains Mars' age, distance from the Sun, and early planetary formation.

Sources: Destination Mars: New Explorations of the Red Planet by Rod Pyle, Prometheus Books, 2012; www.esa.int

NASA spent close to USD 2 billion on its Mars Curiosity mission. In this artist's impression Curiosity is seen being lowered onto the surface of Mars.

Credit: NASA

minds. At the same time, several of their deductions through telescopic observations seemed fairly accurate, and sometimes it seems that we have known a lot about Mars for centuries and that the addition to our knowledge has perhaps not been as much as we would like to believe.

But, it isn't as though everybody was right all the time. Years and years of human study of the Earth's exciting and mysterious neighbour has resulted in a lot of exaggerated interpretations that are both fascinating and somewhat ludicrous at the same time. Back in 1860, an English astronomer conducted telescopic observations and wrote how there was no part of Mars that could not be reached by ship! As Rod Pyle has said: 'It has since been argued that Schiaparelli, Lowell and the others who so patiently charted the lines across the Martian surface may have

A selfie of the Curiosity Mars rover.
Credit: NASA

done little more than traced the capillaries in their own retinas, either as shadows cast through the structure of the eye or as reflected in the eyepieces of their telescopes. Nobody can be sure. What can be said with some authority is that no two observers saw quite the same patterns, and few users of modern telescopes have felt compelled to make note of such patterns in the last one hundred years.' To ancient Indian astronomers, Mars was important, as was Saturn, and they used rudimentary measuring instruments in the absence of telescopes. All these early astronomical studies pieced together a picture of a planet very, very similar to Earth.

So, at the turn of the nineteenth century, came the strong belief that there could be life on Mars. However, although human imagination has always been fertile, it has also been tempered by the exacting human mind that has, in turn, led to an entire body of knowledge that we know to be science today. The two have managed to coexist, although sharp facts and observations that characterise scientific inquiry have generally led the way. That is why the

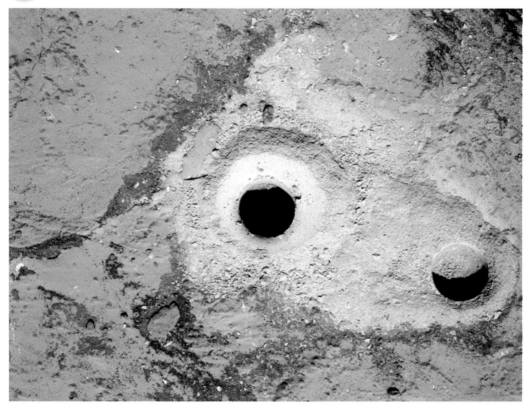

What lies beneath the surface of the red planet? NASA's Curiosity Mars rover drilled this hole on the surface of Mars, but has not found any trace of methane on the planet.

Credit: NASA

US Mariner 4 of the mid-1960s was critical, bringing images the way it did, smashing several earlier astronomical predictions. The Mariner 4 was the fourth in a series of flyby spacecraft missions, and the first successful one that the US sent to Mars. This was the first time that any space exploration managed to capture images of another planet from deep space, and while these images largely painted a picture of Mars as being quite a dead planet, with neither canals nor any sign of life, the search goes on, and the *Mangalyaan* mission that India has mounted is no different. Something obviously fuels this search, despite the red and dead images, the enormous craters and volcanoes, and the many other traits that make up Mars.

Land and icy hilltops, what more do humans want? Under a pink sky exists a cold planet, with an average temperature of 20° Centigrade, dipping to a minimum of −140° Centigrade. That is why the carbon dioxide (CO_2) present there freezes to form a massive polar cap, alternately at each of the poles. This great cover of snow melts in the spring. Then there are layered volcanic rocks, impact craters, cratered plains and dark sand dunes, with the latter raking up rather violent dust storms for extended periods of time through the course of the seasons. In very earthly style, frost and fog are also known to set in seasonally, and there are clouds too. All this makes for conditions that seem eerily similar to those on Earth, with surface features of the planet reminiscent of volcanoes, deserts and polar ice that humans are used to in this world. It also perhaps snows there, with snowflakes[4] made of CO_2, so there is a clear picture of 'weather' and climate on Mars. There are suggestions that a colder or drier climate was prevalent at some time, because clays were formed in the lake environment.[5] So, are Martian seasons, climate and the weather in anyway facilitators that lead to the eternal teaser, as to the possibility of habitation? Not really, if you ask the experts. Sujan Sengupta, who works at the Indian Institute of Astrophysics in Bangalore, says Mars is far from being a habitable place, and there are several reasons why. The extreme nature of each season is one. Despite Mars reflecting as much solar light as the Earth does, its highly elliptical orbit around the Sun creates great temperature differences. Also, the rather high tilt of Mars's rotation axis is quite high and changes rather chaotically because unlike Earth, Mars does not have a massive and influential satellite like the Moon. Then, the large amount of CO_2 is a deterrent, made worse by the action of intense ultra violet rays and charged particles carried through solar wind that break it down into the poisonous carbon monoxide.

Piecing together information gathered through the study of Martian meteorites in the laboratory along with remote sensing data from past missions to Mars, the world has built up a picture of what it is like on the red planet. Observations such as these have not only helped develop a picture of what the Martian surface and environment is like in contemporary times, but also helped with the understanding of what it might have been like much earlier. Doubtless, there is soil and rock, air (of whatever kind and quality) and ice, and a specific geology.[6] The four rocky, terrestrial planets that compose what is called the inner solar system are Mercury, Venus, Earth and Mars. This has been fed by written and pictured accounts, through cinematic extravaganzas that tell stories of aliens having landed, and equally through scientific knowledge of what it is like on Mars. Jupiter and Saturn are gaseous and the last two planets of the Solar System are the icy Uranus and Neptune. Mars has an orbit that moves between the Earth and the asteroid belt. The Moons of Mars are not only small, but also irregular, and the material that seems to compose them is similar to the asteroids in the nearby asteroid belt. This begs the question whether Phobos and Deimos are actually captured asteroids. Phobos moves around Mars in roughly seven and a half hours, Deimos in about 30 hours. On the planet, the Northern Hemisphere is flat and low-lying, compared to the Southern Hemisphere with its highlands and most of the Martian craters. There is no other planet that has volcanoes as large as those on Mars and canyons as deep. That said, it seems the volcanoes are no longer active, although they perhaps were in some earlier epoch, with estimates suggesting that the last lava flows perhaps occurred roughly 20 to 200 million years ago. In fact, speaking of early times, after the original Martian crust got formed through the cooling of a magma ocean, its differentiation into crust, mantle and core happened relatively quickly, within a few tens of millions of years of the formation of the Solar System. The atmosphere on Mars was apparently very thick in the beginning, and has slowly depleted over time by thinning out. In the geological evolution of Mars the first era was 4.5 to 3.5 billion years ago, called the Noachian Period. Violent eruptions occurred during this time and are represented by the massive, overlapping craters that were created then. The Noachian Period seems to have had both alkaline and extremely acidic conditions co-existing, with the alkaline conditions preceding. This was followed by the Hesperian Period 3.5 to 3 billion years ago, during which filled basins got created because of the basaltic magma flowing out of the planet's interior. Finally, it was the turn of the Amazonian Period, from 3 billion years ago till today. In present times, the surface of Mars is neatly classified into some well-mapped regions, because all missions of previous years have made visual and other observations. Mars has the Yellow Knife Bay at the Gale Crater, the famous Valles Marineris which is a 4,000

Unravelling Mars in detail. This lovely, false colour image of Mars shows the presence of craters within craters. Surely, a very hostile environment for the existence of life.

Credit: ESA

kilometres long valley that is roughly along the Equator, the Hellas Planetia, also called the Hellas Impact Basin, an enormous impact crater, in the Southern Hemisphere, and of course, Olympus Mons, which is perhaps the highest volcano in the entire Solar System. In fact, it is apparently three times the height of Mount Everest. Well before India launched its mission to Mars, several craters on the planet were already named after Indian towns, like the Broach crater with a 12 kilometre radius, named after Bharuch, and Poona, named after Pune. Almost 40 per cent of Mars is taken up by the Smooth Borealis basin in the Northern Hemisphere. Martian soil is slightly alkaline and has sodium, potassium and magnesium, with basalt and the dust of Haematite forming the majority of the surface.

Table: Even before India reached Mars, there are craters on the
planet named after Indian towns and cities

1.	Broach	12.0 km	1976
2.	Kakori	29.7 km	1976
3.	Lonar	11.3 km	2007
4.	Poona	20.0 km	1986
5.	Rayadurg	21.8 km	1991
6.	Sandila	13.9 km	1976
7.	Wer	3.1 km	1976
8.	Bhor	6.0 km	1979

Early Mars was habitable, but that does not mean it was inhabited.[7] Even if there is some microbial or nascent life hidden somewhere in the rocks and crevices, it has not managed to oxidise the atmosphere the way cyanobacteria did on Earth. 'I don't believe that Mars ever had an atmosphere favourable for supporting life. Such statements are purely speculative and have no supporting evidence. Mars is similar to Earth only in the sense that it has a rocky surface, an extended atmosphere, similar albedo or reflectivity, and sometimes its temperature goes higher than zero degree Celsius. But it is the dissimilarities that make Mars uninhabitable.' Sengupta is not alone in having this view point, but even so, the most critical search is for signs of life, or shreds of organic matter, anything that would show the remotest possibility that there could be life on Mars. This search for organic matter on Mars is what is preoccupying most people now, as is the search for methane gas that is a by-product of organic matter. The Curiosity rover's search is on and NASA has already conducted three space missions before this (Viking Landers 1 and 2 in 1976, and the Phoenix Lander in 2008) to Mars to detect organic matter on the surface; none of them came back with any confirmations. This missing piece is hard for most experts to comprehend, apparently because it is known that Mars receives about 1,000 tons of organic matter each year in the form of organic rich cosmic dust from comets and

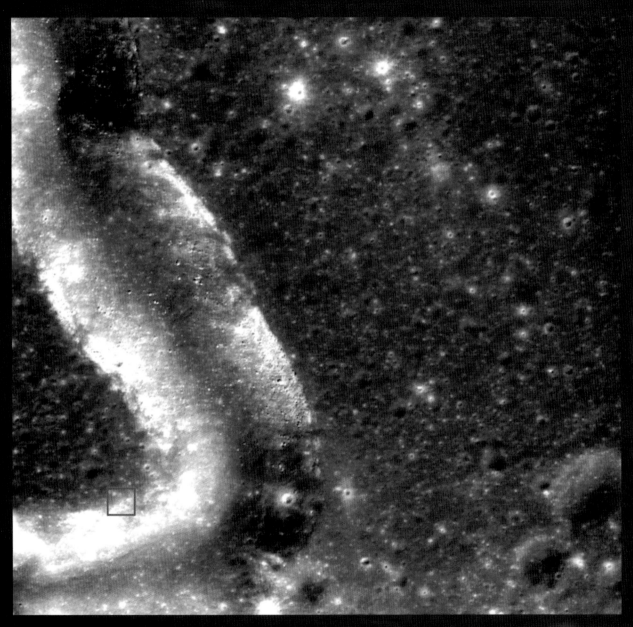

Chandrayaan-1 sets to rest conspiracy theories that America never landed on the Moon. This mission brought back third party evidence that America had indeed landed on the Moon. Close examination of the portion enclosed in the red square in the picture reveals the artefacts left behind on the Moon through the Apollo missions that sent one dozen Americans to the Moon.

Credit: ISRO

asteroids.[8] This has further led to the idea that there might be a chemical that is adding oxygen to organics and turning them into CO_2, giving the chemical the nickname of 'superoxidiser' in Martian soil. This could be a perchlorate, and then there are cosmic rays, and ultraviolet radiation, and all of these can degrade organic matter. The currently thin layer of atmosphere is obviously far less protective than the Earth's atmosphere, which keeps a lot of harmful radiation away from falling on the surface of the Earth. On Mars, the thin atmosphere allows cosmic and solar radiation to hit the surface, making survival of life very difficult. Its chemical composition is simple – CO_2, water, nitrogen, oxygen in very small quantities. Compared to the air humans are used to, with more than 20 per cent oxygen, Martian air has barely one thousandth of the atmospheric pressure we have here on Earth.

Paradoxically, though, Mars has the right conditions for life. It is confirmed that diverse aqueous environments did exist on the surface of Mars, billions of years ago.[9] Scientists speak of evidence that points to how both Mars and Venus have similar Deuterium to Hydrogen ratios, pointing to the fact that both planets had plenty of water at some stage. Chances are that it might have evaporated from Venus while it got ionised and dissociated into hydrogen and oxygen on Mars.[10] There could be a large amount of water stored inside the crust of Mars. But it may not be in liquid form. All this has led to one of the most debated questions amongst experts who know space and study the Universe, what makes us think there is or was life on Mars? And, the logical follow-up question, can we live on Mars in the future? As precision grew and scientific inquiry was built up, it was recognised that water channels that were being observed on Mars were really not there at all, and it seems there is nothing habitable. These so-called water channels, described as precise criss-cross lines across the surface of the planet, were very often thought of as irrigation channels built by intelligent beings there. But the high quality images sent back by American missions have very clearly ruled out the existence of such channels. Whether Mars is geologically active or not is something of an open question even today. Some experts consider it dead, explaining how there is no recycling of minerals. According to others, there still is some action going on. Gully formation just two to three years ago has been captured quite clearly in recent imagery.[11] Meanwhile, the world prepares to solve the puzzle on habitation. It is obvious that 400 years of focused and steadily intensifying attention on a planet like Mars couldn't be mere chance. One major influencer for this attention is that Mars is perhaps the only other planet in the Solar System that can even remotely be a supporter of life, or has been in the past. Or if the mineral deposits there could be of interest and utility to humans. Or that the technological capability to mount lunar missions needs to be extended to deep space and other planets. Whichever way you look at it, unravelling Martian mysteries seems to be a growing passion and therefore hard to ignore, and is not an open and shut case at all. If anything, it is perhaps the search to solve the Goldilocks

Paradox that makes the Earth the perfect place in the Universe for life, with none other like it. As Sengupta puts it, it takes a formidable combination of favourable characteristics to make a habitable environment,[12] and Mars isn't that place just yet.

Notes

1. P. Bagla, Indo-US Space Ties Ready for Take-off: NASA Chief, Interview with Administrator of NASA General Charles Bolden. *The Hindu*, August 29, 2013.

2. S.K. Das, *Destination Mars: Secrets of the Red Planet Revealed*, 2013.

3. http://www.esa.int/About_Us/Welcome_to_ESA/ESA_history/Jean-Dominique_Cassini_Astrology_to_astronomy. Accessed on June 7, 2014.

4. A. Lele, *Mission Mars: India's Quest for the Red Planet*, Springer Briefs in Applied Sciences and Technology, 2014.

5. J.P. Grotzinger, 'Habitability, Taphonomy and the Search for Organic Carbon on Mars', *Science*, vol. 343, January 24, 2014, pp. 386–7.

6. Mars Orbiter Mission, Study Report (extracts) as on July 2011, ISRO.

7. J.P. Grotzinger, op. cit.

8. Richard A. Kerr, in 'The Hunt for the Red Planet's Dirtiest Secret. News Focus', *Science*, vol. 337, August 31, 2012, pp. 1032–3.

9. J.P. Grotzinger, op. cit.

10. S. Sengupta, Personal Communication, June 2014.

11. Mars Orbiter Mission, Study Report, op. cit.

12. P. Bagla, NDTV, Mars is Hostile and Parched for Life: Interview with Professor Sujan Sengupta, October 20, 2013.

Chapter 4

Crafting the Craft: A Hundred Metre Dash for the Martian Marathon and Stories of Frugal Indian Technology

He had almost forgotten what it was like to live a normal day when he would wake up at home, go to work and come back to his family in the evening. The sprawling ISRO Satellite Centre (ISAC) in Vimanapura in eastern suburban Bangalore had become his home for almost 15 months, as he led a charged team of young engineers on what has gone down in Indian space history as the fastest satellite technology development saga – a hundred metre dash for a marathon to Mars! Scientist and Project Director for the Mars mission, the genial, strapping S Arunan describes his challenge: 'We have pushed the boundaries, whether with the ground segment, satellite or launchers. Now we are living the dream.' Indeed.

The exploration of space has always found special place, and in a dreamy, magical way, the Universe always flits about the human head and heart, so near and yet so far sometimes. So, when Arunan says he is now 'living the dream' along with the rest of his young and dynamic team at ISRO, I (Subhadra) wonder whether this is the young boy in him who speaks or the industrious scientist who has worked without stopping for over 15 months to get to this point. After all, Arunan, by his own admission, is a 'satellite manager', a description that seems to deliberately strip his job of its magic, tossed into the mundane. Besides, the nonchalant way in which he says it, half in jest, makes it all sound very elementary. What he actually does is anything but elementary, but then the man's humility is like a second skin: 'I don't know what they saw in me, but I am indeed fortunate to have been entrusted with such a responsibility.' And it is this feeling that drove this amiable man to make the satellite centre at ISRO's headquarters in Bangalore his home through endless nights and days spent fashioning the mission. 'All this time, I would go home only to freshen up and say my prayers.' There is no surprise why stuff like this is what dreams are made of.

It is, nevertheless, only an unfolding event, and one that India has been a late-comer on in terms of technology. But from Arunan's vantage point, that's a plus. 'Failures of other nations are stepping stones to my success', says the man who has handled many satellite missions, describing how detailed reviews of earlier missions have helped. Like so many others at ISRO, Arunan too is now savouring the experience of having crafted the craft that is on an inter-planetary adventure. His wish list for taking on the job was small; 'I requested them to depute youngsters to work with me, say between 25 and 30.' Arunan is a staunch supporter and follower of the young because he sees in them irrepressible enthusiasm, a great desire to learn, and an even greater desire to perform. 'You give them a task and they take it on with so much energy.' As if on cue, and quietly, as we chat over green tea served in modest teacups, a small group of youngsters assembles on the side table. We will later sit together and talk about their efforts with the Facebook page on the Mars mission and how they see that as the next big thing.

I (Pallava) feel the same youthful verve when I am allowed to touch the gold-wrapped object that looks like a Tata Nano car with wings and stands solitary in the sharp gleaming hues of the clean room. When I first saw the satellite at the ISRO Satellite Centre (ISAC), engineers were buzzing around. A lot of testing was going on and the satellite was powered. I wanted to file my television story standing next to the satellite, and when the team took a break they allowed me to go closer. A step ladder was placed and I climbed up. You have to wear a special plastic band reinforced with some copper on your hand to remove static electricity, so that you don't pass it to the satellite. As I touched it, I said spontaneously, 'It smells of India'.

How human and how youthful is this feeling of excitement and anticipation which gets enhanced when the mind's eye conjures up a golden object hurtling through the black emptiness of the Universe. The magic and infectious excitement of space exploration has made many a hard-boiled adult into an inquisitive child, so who am I to be an exception. But the job of creating a satellite is nothing to do with magic, it is all about engineering science, precise and exact, with no margins for error. And with the Indian space programme having launched more than 70 satellites in the last four decades, there is certainly something ISRO's satellite managers are doing right. Also, since all this is happening in a country that defies logic on why it has or needs such an active space programme, M Annadurai, Programme Director of the Mars mission is encouraged to talk about the 'Indian' way of doing things. 'We know we have only one chance, there are no luxuries. In a country like China, they can afford to start planning to host the Olympics three game cycles earlier, here, its different,' he says, chuckling. Annadurai is obviously comfortable with the Indian way. The fact that in India we 'go for one model, and it has to be a success' makes a great difference to the approach that is taken in executing each and every project in space exploration. This means that each experience of the past is used rigorously, whether a success or failure. ISRO had burnt its fingers during the Chandrayaan-1 mission when the satellite got heated up, so these lessons are carefully used in fabricating the MOM satellite. Recent times at ISRO have seen team dynamics alter with a dramatic upward swing in confidence levels after the success of Chandrayaan-1. The MOM satellite is an engineering marvel, built to tolerate extremes in temperature, an unimaginably long and lonely journey, and against a shoe-string budget in record time. As we sit in Annadurai's spacious office and I nibble at a biscuit, he points to a small photo frame placed prominently on his table: 'Failure is not an Option' declares the famous quote. Although reflective of the American space programme, the phrase does seem to capture the Indian spirit too. 'Look at the Jet Propulsion Laboratory, they have already begun working on a 2018 mission,' says Annadurai. The Indian 'way', therefore, may seem like a slap-dash, street-fixer style (what we call *jugaad* in India), but in the final acid test, everything must perform to the exacting demands of a deep-space environment. No shortcuts there.

Meanwhile, in the cavernous clean room at the Space Applications Centre (SAC) in Ahmedabad, the young and vivacious scientist who speaks with her eyes dancing with ebullience infects me with her *joie-de-vivre*. How can she be so thrilled about this small, inconsequential thing? I am thinking quickly to keep pace with her rapid chatter, staring at the little black box that is the centre of her attention and perhaps the source of her exultation! That's the whole thing about being a scientist. The fact that they had assembled this tiny thing in record time, that it would ride the MOM satellite and answer a few dramatic scientific questions of which the answers were so far unknown, is sufficient to bring the sparkle to her eye. JN Goswami, maverick Director of PRL, is happy about these 'simple, straightforward instruments' they have developed, as part of what he succinctly calls a 'comprehensive mission with complementary instruments'.

In the pursuit of science and in discovering new knowledge, it is sometimes hard for those outside the circle to actually understand what's going on. It is equally tough to empathise with or even comprehend the fire in the belly that leads scientists to execute the most inexplicable of projects. That is why they are forever on the lookout for any window of opportunity that opens up, and this is particularly true in India. In India, where accomplishing anything that is even fractionally more sophisticated than building a toilet or providing safe drinking water, getting life-saving vaccinations for babies, or supplying electricity to all citizens is questioned for its necessity of purpose, the pursuit of high knowledge can be a troublesome thing. Of course, there is an irony here; the same basics described are India's biggest bugbear despite large governmental programmes and resources devoted to each of these developmental needs and more, mired as they are in implementation bottlenecks, inefficiency and an alarming lack of accountability. Anyway, the Indian space programme has kept itself going inspite of this frequent and critical refrain, and the Mars mission is special because it not only tests

ISRO Chairman K Radhakrishnan holding a copy of the *Study Report on Mission to Mars.* This June, 2011 report encapsulates results from a feasibility study conducted by ISRO focusing on India undertaking a mission to the planet well before the government gave its formal clearance for the mission.

Credit: Pallava Bagla

mettle in terms of inter-planetary travel technology, but also allows India to ask some of the most popular and persistent Martian mystery questions of this century and the last. Goswami believes that each opportunity to ask questions creates a new wave of enlightenment, and 'unless we gather knowledge, you cannot understand everything in nature. Our understanding of the Moon has changed radically since we found water through Chandrayaan-1, something every other lunar mission had missed.' He makes the point quite unambiguously, that despite a series of lunar missions by other nations being mounted before Chandrayaan-1, the Indian mission did make a spectacular new contribution to the world's understanding of the lunar environment and surface. Therefore, it is highly presumptuous to think that a nation does not need to repeat the space exploration feats of another. In the current geo-political environment where there is hardly any sharing of data or technology, the do-it-yourself principle is what seems to fit the bill.

With the Mars mission, there is even greater excitement because it moves from one planet to another. Routine missions do not give such great opportunities of learning, of sharpening the mind, and its ability to deal with highly niche areas of space engineering, technology and science. As the world debates endlessly about why India should be going into space, the indisputable fact is that India is firmly in space. All over, actually, and the country's first inter-planetary mission seems to cement that firmness and clearly puts the writing on the wall – if Mars is a destination for humankind in the near future, then India is headed there too. After all, the developed nations of the world have proved time and again that there are no free rides for the less privileged. Interestingly, NASA sent ISRO 'lucky peanuts', following a superstition that they would herald a successful mission. India isn't, after all, the only country where superstition is followed.

Once decisions were made and plans laid for the MOM mission, it was a race against the clock to get the satellite ready, which is at the heart of the mission. Pumping this heart is the extensive ground stations network that must keep track of the satellite at all times, and attributing greater meaning to this entire effort is the payloads, or scientific instruments. Needless to say, the rocketry that forms the backbone of all spatial adventures is a key pillar, and its interesting details are covered in Chapter 5. All in all, despite being a technology demonstrator, the mission also has some clear scientific objectives, with deeper purpose. In the development of these four main pillars, the uniqueness has been shaped by the fact that India was flying solo this time around, going far, and had little time to tick all the boxes before the flight in November 2013.

A detailed mission sequence was worked out keeping October 22, 2013 as the launch date in mind. ISRO always had the Moon, Mars and the Sun on its mind, therefore it was between

Lucky Peanuts

As you prepare for your launch to Mars, do not forget one of the few, but important actions: pass around the peanuts!

The tradition of the peanuts for Mars missions at JPL goes way back to the 1960s with the very first missions we sent to the moon. We had seven mission attempts to go the moon before we succeeded, and on that seventh one, they had passed out peanuts in the control room. Ranger 7, which in July 1964 became the first U.S. space probe to successfully transmit close images of the moon's surface back to Earth, made the peanuts into a tradition at JPL. So ever since then, it has been a long standing tradition to hand out peanuts whenever we launch and whenever we do anything important like land on Mars. We use all the luck we can get!

For MSL, we put a label on the jar that says "dare mighty things". That phrase was taken from Theodore Roosevelt's quote, "Far better it is to dare mighty things, to win glorious triumphs even though checkered by failure, than to rank with those timid spirits who neither enjoy nor suffer much because they live in the grey twilight that knows neither victory nor defeat."

GO MOM!!!!!!!

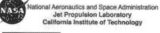

National Aeronautics and Space Administration
Jet Propulsion Laboratory
California Institute of Technology

GOOD LUCK MOM!
LAUNCH 5 November 2013
DARE MIGHTY THINGS!
(Mars peanuts for launch success!)

Lucky peanuts: superstition or mission requirement? India launched its Mangalyaan on *Mangalvaar* which is Tuesday in Hindi, and mangal means 'auspicious'. The world's foremost space agency NASA it seems is equally superstitious as is evident from this story of how passing peanuts around has become something of a lucky charm. Scientists of the Jet Propulsion Laboratory of NASA sent a good luck card to counterparts at ISRO along with a bottle of peanuts.

Credit: ISRO

August 2010 and July 2011, a team led by Adimurthy, Senior Advisor on Inter-planetary missions at ISRO, came up with the feasibility study regarding the mission to Mars, as mentioned earlier. In 2012, individual components of the Orbiter began to get assembled, and they were all pulled together in March 2013. Come April and instruments began to get integrated after final tests and reviews. It was only in September that the mission was unveiled to the media. The finally-tested satellite with all payloads integrated went through tests and checks at the sprawling ISRO Satellite Integration and Test Establishment (I-SITE) in Bangalore, at the huge Marathahalli campus, and was moved to Sriharikota by road in October. There was a minor postponement of the final launch because the tracking ships were not in place at that point of time, owing to rough weather out at sea. The ideal time for launch would have been October, because it would have meant avoiding the cyclone season, but storms in the south Pacific Ocean pushed it to November. Then, one of the ships also developed a problem with its generator. Finally, November 5, 2013 dawned and the future became the 'now'. The electricity of the moment was palpable, as I stood on the famous Launch Pad 1 at Sriharikota, knowing as I did that seldom is a person allowed to stand so close to a fully armed and powered rocket. What thrill there was, in interviewing the ISRO Chairman Radhakrishnan, standing right there. Radhakrishnan gave me several interviews, but this was no doubt special and extremely thrilling. 'We are ready for a historic launch,' he said. With a small rocket, and a small satellite, were the dreams and ambitions too big? But Radhakrishnan is incredibly confident about the PSLV-XL. During the final run-up, all attention of the ISRO teams was focused on spotting areas where there was some risk to the mission. So, certainly not butterflies in the stomach, but a careful eagle eye for averting any crisis.

For the designers of the mission a few big questions loomed large. First of all, going by the admission that it is primarily a technology demonstrator mission, close to 85 per cent of all weightage went for getting the technology right, first to reach and then orbit Mars. Charting long distances and therefore figuring out how the satellite would be powered, ensuring long-term propulsion, cracking the communication poser of how signals could be delivered over hundreds of millions of kilometres, and orbiting Mars finally – each of these is a colossal mission in itself, with a series of smaller challenges of its own. Coming a humble second therefore, was the question of conducting meaningful experiments during the lifetime of the satellite. The fact that the PSLV was the vehicle of choice, although this could be less about choice and more about availability, placed a significant constraint on payloads. This, in turn, necessitated some sharp and smart thinking in framing the right scientific questions. But first and foremost, the satellite had to be engineered into existence.

The satellite bus arrives at ISRO. Seen in the picture are officials from the Hindustan Aeronautics Limited formally delivering the primary satellite structure to ISRO's team, led by M Annadurai, Programme Director, in September 2012.

Credit: Hindustan Aeronautics Limited

CRAFTING THE CRAFT

On April 9, 2014, five months after the launch, the MOM satellite crossed the half-way mark on its long journey to Mars. This was doubtless a milestone considering the distance covered was 337.5 million kilometres. Just a few months before, in the wee hours of 1 December, 2013, a tricky manoeuver called the trans-Mars injection was carried out, which meant that the spacecraft had moved out of the Earth's orbit, heading towards Mars, after getting successfully placed in a Mars transfer trajectory. An ISRO press release at the time announced, 'The critical manoeuvre to place India's Mars Orbiter Spacecraft in the Mars Transfer Trajectory was successfully carried out in the early hours of today (Sunday, December 1, 2013). During this manoeuvre, which began at 00:49 today, the spacecraft's 440 Newton liquid engine was fired for about 22

Burning the midnight oil to reach the red planet. Satellite engineers beavering away at the Mangalyaan fabrication facility at the ISRO Satellite Centre (ISAC) in Bangalore.

Credit: Pallava Bagla

minutes providing a velocity increment of 648 metre/second to the spacecraft. Following the completion of this manoeuvre, the Earth orbiting phase of the spacecraft ended. The spacecraft is now on course to encounter Mars after a journey of about 10 months around the Sun.' Jubilation it is, nothing less! To our satellite manager, Arunan, these are major milestones. Although scientists and engineers as busy as these men and women at ISRO have little time for nostalgia, they do like to reminisce about the days when the satellite was getting built. Some fundamentals were non-negotiable for Arunan and his team, that the ground segment and the launch vehicle had to be absolutely ready. 'Once the fundamentals were clear, and I knew we could use the long coasting possibility of the rocket to get into the tricky Mars orbit, I knew we could go ahead,' says Arunan who felt that his primary responsibility would be to 'go ahead and make a satellite worthy of going to a planet, for a launcher can only do this much'. Now the story of shaping the satellite began, and it is interesting indeed to discover that satellite management is certainly not for the weak-hearted.

The MOM spacecraft is a box-like satellite, small by any standards (see Annexure 2). Its design is a hybrid of the kind of satellite bus used for Indian Remote Sensing Satellites (IRS), the Indian Navigation Satellites (INSAT) and Chandrayaan-1. The Hindustan Aeronautics Limited (HAL), Bangalore, established dedicated facilities for ISRO and fabricated the satellite's primary structure, equipment panels and special brackets,[1] delivering them to ISRO in September 2012. Needless to say, Mangalyaan needed a lot of creative improvisation, given the communication, power, propulsion and autonomy challenges. With composite and metallic honeycomb sandwich panels and a central composite cylinder for all the equipment,[2] the actual satellite weighs just 475 kgs, the payloads are 15 kgs, and fuel, 852 kgs. Another major factor defining the creation was the fact that it needed to go really far away from the Earth, therefore requiring a lot of power. The solar panels that are so typical of any imagery of satellites are critical. 'Will they generate sufficient power in very cold temperatures, in the conditions surrounding Mars?' worries Arunan. In designing the power systems one of the major challenges is the distance between the Sun and Mars, and the latter is the last planet in the solar system where solar power generation can be used effectively, although solar irradiation there is half of what it is on Earth.[3] That is why Mangalyaan has three solar panels compared to Chandrayaan-1, which had one. The satellite's thermal systems have to maintain the spacecraft at about 20 degrees Centigrade inside. Meanwhile, having fuel on board is an issue because of the weight constraint. The main propulsion system of the satellite is driven by a 440 Newton liquid-fuelled rocket engine. The satellite is powered using its on-board rocket motor and 850 kgs of fuel stored within, which imparted enough velocity for it to escape the Earth's influence and take off towards Mars in December 2013.

Moreover, the cost has been a steal. The satellite alone has cost just ₹ 1,500 million, and even the entire mission just ₹ 4,500 million (which is USD 76 million when the rupee rate was 59 to a US dollar), compared to the massive USD 671 million spent by the US's NASA on Maven, which is the US mission to Mars, very similar to India's MOM and launched in 2013. It is interesting that NASA, for many years, had a 'faster, better, cheaper' motto, which of course lost favour after the country's twin Mars mission failures in 1999. In India, shoe-string budgets are a day-to-day reality, and over the years, ISRO has learnt how to work within these and yet deliver quality. The MOM spacecraft was built in one shot, ready for flight. The final configuration required for actual flight, also called the final flight model, was the one and only model built, right from the start, and using the latecomer advantage, India also restricted the option of going for too many ground tests. On several components, there are no compromises on budgets of course, like the star sensors on the satellite that have to be bought, whatever the price. Many innovative and specific actions had to be taken to stay within the budget. Another related aspect that added stringency was the gruelling timeline for delivering the satellite at Sriharikota, and here again there is a compelling comparison – NASA's Maven was six and a half years in the making, ISRO's Mangalyaan was up and ready in 15 months flat!

Ruggedness of the satellite was a major priority for the designers. Considering the conservative flying approach that the Indians had decided to adopt, there was a 20–25 day period during which the spacecraft would repeatedly cross through two belts surrounding the Earth, high radiation zones with high energy charged particles, called the Van Allen belts, and these crossings could possibly affect the on-board electronics. Designers, therefore, needed to take extra precautions, as they had to make the craft rugged enough to tolerate the many adverse conditions of deep space. Although lightning strikes are more common nearer the Earth, an asteroid hitting the craft could be a remote possibility, but a possibility nevertheless.

For experienced old-hands like Arunan, there aren't too many surprises in satellite creation or management, but for Mangalyaan, teams of scientists and engineers wrestled with a new piece – on-board autonomy. The spacecraft had to be made independent so that it could take key decisions in order to stay on course, perform its many functions and complete a successful run. 'Our first question was to think, what are the ways in which the system can goof up?' is Arunan's description of how the team took on this new technological challenge. All combinations of failure and success had to be simulated and the spacecraft had to be made worthy of identifying and overcoming glitches without ground intervention. Why? Simply because the ground stations in Bangalore are just too far away and unreachable, and the last thing ISRO wanted to hear from the MOM spacecraft is 'Bangalore, we have a problem'! This threw a dare at the engineering design because in normal course, ground controllers

ADVANCED MISSION COMPUTER

Powered by silicon. The advanced mission computer for Indian space missions.
Credit: Pallava Bagla

of a mission would keep a strict eye on all systems and their performance, besides giving commands from the ground as required. Here the spacecraft is on its own, and impervious to real-time interventions. That is why it is designed to use a system called Fault Detection, Isolation and Reconfiguration (FDIR) to check on all systems and their performance.[4] The idea is simple, but the technology complex – an isolated machine somewhere out in space, working to detect, isolate and reconfigure, and to keep going. It will be much later that the ground systems and scientists will come to know of any reconfiguration that might have happened, why it was done, and also intervene if the satellite happened to take any wrong decisions. The satellite's own mechanisms must take it where it wants to go and perform well, after the rocket does the initial bit.

Such full-scale on-board autonomy is a first for ISRO, and that is why the MOM spacecraft had to undergo extensive testing without any compromise. The distances involved are so large that a signal from the ground to the craft or the craft to ground can take anywhere between six to over 20 minutes one way. By the time ground stations come to know of a problem and communicate a solution, the satellite would have gone somewhere else. Autonomy is equally critical if the payloads have to operate smoothly and perform functions they are expected to,

MENU

Kim-Chi with 2 Kinds of Side Dish
백 김치와 2종 찬 (오징어 젓, 소고기 장조림)

Jelly Fish with Shrimp & Mustard Sauce
겨자소스를 곁들인 해파리 냉채와 새우

Milk Porridge
타락 죽

3 Kinds Of Jeon
삼색전죽 (애 호박, 관자 전, 표고)

Char Grilled Rib-eye Steak & Dry Fig Sauce With Grilled Vegetable
& Bamboo Leaf Rice
무화과 소스를 곁들인 그릴에 구운 꽃 등심구이와
구운 채소들과 대나무 잎 밥

Permission Pudding With Seasonal Fruits
감 푸딩과 계절과일

Green Tea or Coffee
녹차 혹은 커피

AMORIS

Banquet & Convention

because each needs a complex chain of commands that will help orient it and trigger it for action. The spacecraft has all commands stored, triggered only when an on-board system gives a command. All this is pre-programmed of course. Finally, what if something major goes wrong? The autonomy that is built into the satellite will nudge it into what is called a safe mode which can allow the spacecraft to then wait for correct commands to be received from the ground. In safe mode, the satellite is so positioned that its antenna points towards the Earth and the solar panels should be in a position to receive the Sun's energy. At this point, it can receive commands. All these systems, including the three levels of autonomy, had to be tested and re-tested for the satellite team to finally be confident and also have the final trigger in their hands – what if MOM were to go berserk and if the satellite were to misuse its independence and get lost, the ground system would have the provision to over-ride all on-board commands.

So much depends on the testing phase. It is easy to feel the tense anticipation mixed with anxiety that the project managers feel when they conduct these tests. As it is, the large, theatrical room at I-SITE in Bangalore has a certain melodrama about it. Covered all over with cones of grey-blue foam, this enormous room is where the satellite is tested in an environment

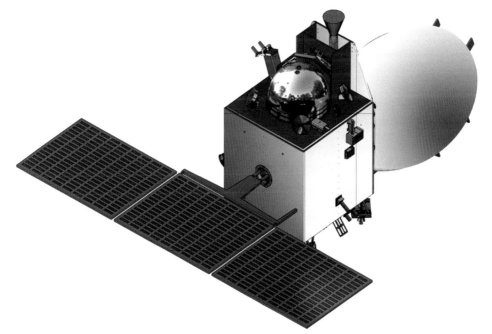

An artist's impression of the fully-deployed Mangalyaan, or the Mars Orbiter Mission.
Credit: ISRO

The hundred metre dash for the Martian marathon. Satellite fabrication activities in full swing at ISRO in Bangalore. ISRO made Mangalyaan in a record 15 months, the fastest satellite development that the Indian space agency has ever achieved.

Credit: Pallava Bagla

simulating conditions of deep space, and free of any electromagnetic interference. This is the Comprehensive Antenna Testing Facility, and standing there it is easy to feel transported to another world in a strangely silent reverence. DR Suma, the lively and energetic woman who is General Manager of I-SITE, does not think it is special though, since it is just course-of-the-day stuff for her. Once out in the black empty spaces, a satellite faces solar radiation pressures from all the dynamic forces that operate out of the Sun, not to mention the other unknown and unforeseen dynamics of the Universe itself, so all the ground work that goes into shaping and preparing the satellite for its adventure is critical. The satellite bus has to cope with near-Earth conditions of heat and the cold near Mars. Save for a few components imported from outside, the entire design and building and testing of the satellite is Indian, and the pride in Radhakrishnan's voice is clearly discernible when he says: 'The level of technology that has gone into this spacecraft, specially the autonomy that has been built in is excellent, and the spin-off of this would be useful tomorrow for our own communication satellites. That means there is technology upgradation taking place on those satellites too.'[5]

The actual day of the launch was intriguing, to say the least. After everything had gone off to precision, media had gathered and the traditional, post-launch press conference was to start. A bunch of local journalists chose the occasion to shout down the scientists and launched a protest against the creation of the new state of Telengana. Holding placards that said 'Don't bifurcate Andhra' and 'No to Telengana', these people took the ISRO establishment completely by surprise. For a good 10–15 minutes after the mission, proceedings at the press conference were overpowered by this shout-down with flummoxed officials incapable of using force, or of keeping silent. All of this was unfolding in the glare of the media gathered to attend the Mars Orbiter launch announcement. As I sat in the audience, this was going on, and it was all rather frustrating. I was thinking how the journey to Mars had begun on a good note, but earthly protests for the creation of India's twenty-ninth state were playing party pooper! Obviously, ISRO and their marshals could not do much, but, using my loud and booming voice I had to shout down the protesters by saying 'silence'. I said it three times. Since I was also one of them, it seemed to have an impact. The protesters melted away. Through the rest of the day, many people from ISRO thanked me, and I noticed how, in the middle of the press conference, Radhakrishnan connected with me silently and expressed his thanks. That same evening, Radhakrishnan decided to return earlier than planned, first to Chennai and then Bangalore, and said it was because of the likely protests against the creation of Teleganga. Just before leaving, he gave me the only one-to-one post-launch interview and also deputed Kiran Kumar, Director, SAC, to do the live show for New Delhi Television (NDTV). Suddenly, around 7.30 p.m., Kumar said that he too wanted to go back to Bangalore. Realising how on the day of a successful mission there would be no senior person left at Sriharikota, I almost had

Martian glory eclipsed by protests. The post-launch press conference by the Chairman of ISRO saw an unruly bunch of local Andhra journalists staging a protest against the then proposed bifurcation of Andhra Pradesh. ISRO's officialdom was caught off guard and left dumb-founded.

Credit: Pallava Bagla

to jump in front of his car to stop him. Taking it all a step further, I threw a question: 'Who should I get for the show then? Would it better if I got Dr Madhavan Nair?' I must admit it was a naughty question, to even suggest that I get the former ISRO chief who is openly critical of the Mars mission, to speak, if Kumar was not available! Whatever it is, this did the trick and Kumar stayed on.

THE ORBIT IS THE BIG THING ON AN INTER-PLANETARY MISSION

If India had the Geosynchronous Satellite Launch Vehicle (GSLV) Mk II ready to take the MOM spacecraft on its journey, things would have been different, in the sense that the satellite circling the Earth orbit might have been avoided. With greater power and a host of other special characteristics (see Chapter 5), the GSLV-Mk II would have had the ability to transport the satellite much farther in one shot. But the priority, as Radhakrishnan said on several occasions, was to operationalise the mission without further delay. A lot of the planning was centred around the orbit, and understandably so, because the ultimate goal of any space mission is to find the right parking space out there.

Perigee and Apogee, words that almost sound like the names of a pair of pets in some free-spirited household! Actually, this is just rocket science jargon to signify the two extreme points in the satellite's elliptical orbit around the Earth. Periapsis is the point nearest to Mars at 377 km and apoapsis is the point that is the farthest, at 80,000 km! All this implies that reaching the Martian orbit is a very tricky job while using a small rocket. Getting to a point to leave Earth and head out to Mars is something a PSLV rocket cannot achieve, because it does not have enough boosting power to push a spacecraft directly onto the path to Mars. But India managed such a feat successfully. The success was because of yet another decision taken again in *desi*, Indian style, to circle around the Earth for a while and then, once velocity had built up, to take the next step, which was the trans-Mars injection of December 2013. 'If there was the GLSV-Mk II we may have avoided circling, but I would still go with circling around the Earth as the preferred option. The Russian Phobos Grunt went for direct injection, and was not successful,' Arunan explains. This is the popular opinion at ISRO, and many of the scientists felt the same way, that being around the Earth for a substantial amount of time is very important. In an environment like outer space, there is the gravity of the Earth, and the Sun's gravitational pull is always there and has the maximum gravity in the entire solar system, governing all planets. Arunan is fond of saying how 'the satellite is but a dot for the Sun in the entire picture.' After some time, as the journey continues, Mars will also be in the picture. In the immediate plan of action the spacecraft has to leave the Earth. Interestingly, although the Sun's gravity is so large, in the 10 lakh kilometres of the Earth's influence, it has no role to play. How much to boost then becomes the question to ask.

Finally, getting into the right orbit for Mars would be a huge technical challenge. For a period of seven months after leaving the Earth's Sphere of Influence (SOI), the MOM spacecraft would cruise along, a period during which several checks and calibrations would be carried out. Overall, there was general comfort in moving around the Earth for a bit. It came with its challenges of course, for instance the satellite would require an immense thrust to move beyond. The MOM was designed to encounter Mars tangentially to its orbit around the Sun, and would go around the red planet once in three days.

It is the first time that an Indian spacecraft is moving out of the Earth's gravitational field. On an inter-planetary mission, a lot of things are very different. Although the fundamental technology is the same, dynamics in terms of trajectory and other such details are quite at variance. So, both reaching Mars and orbiting it for nine months – trans-injection and orbit insertion – are important (see Annexures 3 and 4).

Mating the satellite with the rocket. A few weeks before the launch the satellite was brought to Sriharikota where it was placed on top of the PSLV.

Credit: ISRO

KEEPING TRACK AND MANAGING DATA

If you drive out of Bangalore on the highway to Mysore, chances are you will stop by to see the tourism hot spot, the Big Banyan tree, a 400-year old tree that spreads across nearly three acres. On my way to Byalalu, which is about an hour's drive out of Bangalore, ambling around the banyan and walking among its many roots and branches that loop and dip all over the quaint, enclosed garden that surrounds the tree, I suddenly spot a gleaming structure in the distance. This is the large dish antenna of Byalalu that is part of the Indian Deep Space Network (IDSN). Designed and put up by the ISRO Telemetry, Tracking and Command Network (ISTRAC) in Bangalore at Byalalu for Chandryaan-1, with two Deep Space Network (DSN) antennae, one 18 metre in diameter and the other 32 metre, this set-up emerges in the middle of nowhere like a scene out of a science fiction movie. Amid the granite outcrops so typical of the Bangalore landscape, this 32 metre dish is what is called a Wave Guide Antenna, weighing 300 ton and perched on top of a nondescript and small building that seems more like a room really. Both science data collection and telemetry tracking will be the job of the large antenna. The dish is designed to receive extremely weak signals from satellites millions

Deciphering Martian whispers. At the ranch-like location set amidst dramatic granite outcrops, reminiscent of the iconic Hindi film *Sholay*. India's largest dish antenna with a diameter of 32 metre is used to talk to and send commands to Mangalyaan. Created at Byalalu on the outskirts of Bangalore, this antenna complex first came up to service India's Moon mission and is a part of the Indian Deep Space Network.

Credit: Pallava Bagla

The impressive Mission Control Centre at India's rocket port Sriharikota, a tiny island off the eastern coast of India.

Credit: Pallava Bagla

Mission controllers making final checks before the launch of Mangalyaan on November 5, 2013.

Credit: Pallava Bagla

of kilometres away. The multiple components fitted into the room below take care of the flow after the signals come in.

Meanwhile, the main ISTRAC complex in Peenya, in Bangalore, is a sight for sore eyes. With a building shaped like a flying saucer and perfectly kept, manicured gardens, this is the place that houses the main Mission Operations Control (MOX) for the MOM and here, engineers and scientists are on a round-the-clock vigil. Inside this impressive facility at ISTRAC, where MOM is kept track of in real time, there are at least four or five people in the room at all times, watching over the parameters that tell stories of the health of the satellite. Data flows in copious amounts, and signals flash on giant computer screens all the time. A deep space mission is all about navigation. Knowing the position and velocity of the spacecraft at every instant is most important because of the long distances involved and the communications challenge. It goes without saying that losing track of the satellite is unacceptable. On a daily basis, the ISTRAC team monitors progress of the mission, in multiple ways. Arunan, for example, can sit and monitor this at his office desk, or even from home, through simple SMS text messages on a

Defying the cliché that Mars is for men, Indian women engineers keeping a vigil in the control room at the launch pad in Sriharikota.

Credit: Pallava Bagla

rudimentary cell phone that he carries around. This integrated system of computers, storage, data networks and control rooms support both the launch and orbit phases. The Indian Space Science Data Centre (ISSDC) at Byalalu is where data will be ingested, using the enormous antennae along with their reflectors and sub-reflectors. The ISSDC is a primary interface to collect, process and distribute data to the end-user, while also developing an archiving system.

ISRO had approached NASA to identify and locate ground segment tracking stations in other parts of the world, which are essential because of the journey being long and inter-planetary. A minimum of two ground stations were required, one each in the Eastern and Western Hemispheres, so that MOM is visible even when the Earth rotates away from sight. NASA, also sending a mission at the same time, had to consider this request carefully. Finally, at a hiring cost of ₹700 million, ISRO was offered the Deep Space Network of the Jet Propulsion Laboratory at Canberra, Goldstone and Madrid. Meanwhile, in order to provide global coverage, another first-ever decision was taken by ISRO and the two ships, Nalanda and Yamuna, were deployed in the south Pacific Ocean some 3,000 nautical miles (over 5,500 kilometre) from Fiji to serve as ship-based tracking platforms and S-Band sea-borne terminals. The mission network includes ground stations at Port Blair in the Andaman and Nicobar Islands, Brunei, Biak in Indonesia and Mauritius, and also in Bangalore, Lucknow, Sriharikota

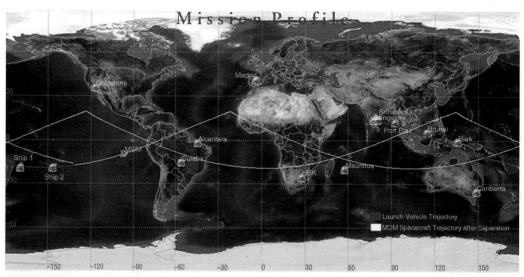

The first few hours of the Mars Orbiter Mission, soon after its launch from Sriharikota. Marked on this world map are the locations of the ground stations that kept track of the health of the spacecraft, including the two ship-borne terminals that India used for the very first time.

Credit: ISRO

and Thiruvananthapuram. Alcantara and Cuiaba belonging to Brazil's National Institute of Space Research, called the Instituto Nacional de Pesquisas Espaciais (INPE), also offer their antennae networks. For a long journey, like the one to Mars, the creation of such an extensive and innovative ground segment tracking design is hardly a surprise.

ASKING THE RIGHT QUESTIONS: MOM's PAYLOADS

If wishes were horses, then the GSLV-Mk II would have helped carry bigger (and perhaps better) payloads for the Mars mission. 'A small payload means small science' is the dramatic declaration by UR Rao who has also chaired ADCOS that took all the critical decisions in determining the scientific aspects of the Mars mission. Meanwhile, Goswami said on a BBC programme that big luggage doesn't necessarily mean it contains very useful stuff. Arunan too believes that carrying less weight in terms of scientific payloads was not a limitation at all. Ideally, the best scientific knowledge gets generated when the research question is first articulated based on a certain gap in our knowledge about something that is of significance. Here, on the Mars mission, it seems to the critics that research questions and payloads have been designed based on how much place there is on the spacecraft. That is not science, according to the purists. Anyway, with ISRO's multi-review mechanisms, what emerges eventually as the final decision becomes gospel, making irrelevant and redundant the many points of view that echo around it. Fact is, ADCOS took the decision to fly five payloads. As pointed out earlier, ADCOS started its work by inviting and reviewing 25 proposals for scientific instruments that would be designed to ask specific questions of scientific enquiry, each meant to further the extent of human knowledge about Mars and its many characteristics. Nine proposals were short-listed, and five finally approved for the flight.

There was a lot of feeling about how very little was being taken on board, as Rao says. There is also a feeling that some of the scientific objectives might have accommodated international collaboration, therefore benefiting from a broader base of enquiry, but none of this fitted into ISRO's race against time on the Mars mission. For the man behind the final decisions of ADCOS, Rao certainly has several reservations. 'Purely from scientific interest, I think we should have gone to Mercury, because there were already quite a few Mars missions. Almost half of these have failed, the bulk of them just before reaching the inter-planetary trajectory.' On the contrary, there has only been one mission to Mercury. So, in more ways than one, the Indian Mars mission is certainly very special, and not at all straightforward. There are many unanswered questions, and the most dramatic of all is that we haven't found life elsewhere in the Universe, nor have we found any Earth-like planets either. It isn't easy to be like the Earth, with just the right distance from the Sun, the right weight, temperature, atmosphere,

and countless other attributes. 'We have not left the baby cot in terms of answering scientific questions,' says Rao, alluding to how much remains to be done. Meanwhile, technology development is taking place all the time, and there is incessant focus on it.

In the contemporary world, human knowledge is now transitioning from 'is there life on Mars question' to 'can we populate Mars?' The pressure of population increase, the ever-growing lack of almost all resources, and shrinking spaces are making people think whether the colonisation of Mars by humans is inevitable. Like Rao is fond of saying, 'It has to happen sometime, and we know what to do. Already, one way tickets are being issued, and more than 1,500 Indians have paid up. When I talk to politicians I request them to think about how the cost of a decent plot of land in Bangalore is today ₹ 25 million!' Is this then a real estate question, one wonders. Anyway, it also leads to the revelation that there is pressing need to do worthwhile science because what we know is so little, and what we don't know is so large. This pull and tug between science and technology is a daily reality for ISRO, and there is no doubt that although deploying scientific instruments is a secondary aim of the mission, and it is not a science mission like Chandrayaan-1, the real excitement nestles in finding answers to scientific puzzles. As a result, at the end of the day there was a clear mandate from ADCOS and the trip to Mars has its fair share of scientific enquiry. The most impactful of all questions is also the most commonplace – is there methane on Mars? This is important because the presence of life can lead to the presence of methane. This is despite the fact that methane can be generated through geological processes also, besides biological. If methane is there, then what produced it? If it is from a biological source, then, as Goswami dramatically questions, 'Are we alone in the Universe? This is a very tough question because we know we have been here only for a couple of million years, we have already evolved for more than four billion years, and we are thinking if somebody else may come. There are so many things, if life or organic molecules developed on more than one planet, in the solar system.'[6] Goswami is also quick to add that there is no simplistic relationship between finding methane and linking that to looking for life because 'life is not a molecule of methane.'[7] One would think not. That is perhaps why several other questions tease the ISRO teams – what Martian atmosphere is like, delving into the records of water on the surface of Mars, understanding mineral composition and mapping the surface of the planet. Mapping Martian polar ice caps and their seasonal variations, monitoring the famous Dust Devils, Martian dust storms, and their dynamic behaviour over six months.

If colonising Mars becomes a serious possibility, all these findings would become critical precursors to such decisions. Somehow, while entering SAC, it is easy to forget this drama. Like countless other government institutions in the country, the buildings are indistinctive, and the early morning queue of staffers clocking in for work is commonplace too. But small things set it apart, the armed guards behind piled up military sand-bags, watching each passer-by with a hawk eye. Saji Kuriakose, Deputy Director, speaks of how back in 2011, they came to know that three instruments from SAC were short-listed for a final place on the satellite. There was hardly any time to waste and no leeway because the launch was fixed. The race had begun, and as detailed earlier, what would in normal course be a four to five year payload development cycle was telescoped into hardly any time in comparison. As many of the scientists and engineers at SAC say, they were compelled to do things differently this time. In the designing of a payload, the most significant factor is the components that it requires, and this is normally fabricated with deliberation and specifications, custom-built after formally releasing Requests for Proposals (RFPs) and characterised by multiple interactions between selected manufacturers and the SAC scientists. This time around, there was no time. Somya Sarkar, the SAC Associate Project Director, initiated an intensive search along with his team of project managers. 'We looked hard at what we already had with us, so the instruments had to be designed around that; second we also picked up what we call Commercially Off-the-Shelf components, COTS for short,' he explains, and goes on to describe how each was then suitably ruggedised to fit the bill. In normal course, Sarkar and others follow a set procedure of first developing their requirements and then carefully fabricating the payload to fit those requirements. Each payload was created by a dedicated team, so there were three teams, seven to eight project managers and some people common to all three teams, 30–35 people in all. Outside SAC, there were many more who were contributing to the purpose of developing these payloads. The *jugaad* aside, there are some unwritten principles that are non-negotiables, and these relate to quality and how best a payload can deliver on its promise. There are no compromises, and that is why the one and a half year period was a whirlwind of activity for the SAC teams, whichever the instrument they were working on. What led them through were the visible and transparent reviews, three or four conducted each month, for each design. Also, project management was very efficient.

For every payload, some factors that influenced the design were common, that each needed to be light, or feather-light. Low power consumption, so it can last a long journey was an important factor too. 'The weight issue was absolutely critical and it was really looming heavily over our heads. These guiding factors propelled us along our journey,' says Sarkar who looks back with surprising fondness at that stressful time. With uncompromising precision, the teams had to ensure reliability, and their way of doing it was to develop three models of each payload, a verification model which was quickly developed by April 2012 for all the three instruments, and this was more like a proof of concept. Then, a Flight Model Like version was readied, which got done around October-November in 2012, and then the final

instrument, ready for the satellite. So nine instruments were developed in total and after final Pre-Shipment Review (PSR), the payloads were packed off to Bangalore.

Knowing what we do know about Mars, the scientific instruments would clearly focus on a few key aspects. For every instrument designer it was important to understand and be attuned to the exact mission objective. The driving force for this quest is not always to search for new stuff, but also to deepen knowledge about what we already know somewhat. The need to look for water and methane is acknowledged. As Rao says, 'If Curiosity has not found it, it is not there in that place, doesn't mean it is not there at all.' Since methane is probably temporal, and geological or biological, it is a most difficult task to find it. There is also the very obvious need for a camera, the MCC in this case. Then, the Methane Sensor for Mars (MSM), its function made obvious by its name, a Thermal Imaging Infrared Spectrometer (TIS) to map the Martian surface and mineral composition, the Lyman Alpha Photometer (LAP) for measuring how much deuterium and hydrogen are present in what ratio, and MENCA for unpacking atmospheric composition. All in all, the five payloads were designed to build on three key themes of scientific inquiry – studies of the atmosphere, environment, and the surface (see Annexure 5). The exploration is fuelled by simple and straightforward instruments that ISRO scientists believe are totally appropriate and will help the organisation piece together a good, composite picture of what Mars is like. In comparison, the space science that NASA conducts is far more dedicated to single points of enquiry on each mission, perhaps because the country can afford to do so. For example, NASA is planning an entire Moon mission just to look for Moon dust. One ISRO scientist likes to believe that the MOM is a comprehensive mission with the complementary nature of the instruments being flown on the spacecraft.

Looking like an ordinary shoe-box, and weighing just about 1.3 kgs, the MCC obviously has to take good pictures. The lightest camera in the history of all those that have been flown to Mars, the MCC is a digital camera and the only one on board, which is why the Principal Investigator for the MCC project, Ashutosh Arya, fondly refers to it as the eye of the entire mission. The camera, according to Arya, does not have any specific scientific objectives, nor does it claim to be the first of its kind. The images will be used to contextualise information from other instruments, grab geo-morphological and structural features that will reveal tectonic activity of the past, capture dynamic events of Mars like dust storms, and, while viewing the poles it will check for ice cover. In Arya's mind, there is something else that is very important. Considering that it is the camera's pictures that can give a first-hand feel to the general public of what is really going on, and what space can be like, he and his team wanted to ensure that the first picture beamed back by the MCC be the one that includes India. At the end of the day, space missions have a great emotional appeal for many people, who feel a lot of national pride in such activities. So, almost as though there was a solemnity about the occasion, the first image from the MCC had a lot going for it. It was planned to ensure that the picture would be taken from an altitude such that it would capture most of India, with the right light and Sun angle, keeping the right orientation of the satellite itself, and also trying to make sure that the shot was cloud free. All of this was figured out, the picture was taken on November 19, 2013 at 0820 Greenwich Mean Time. Showing the Thar desert, fog, peninsular India, and the cyclone Helen building up on the eastern coast, and received by midnight, the picture got 20,000 likes in one and a half days on the mission's Facebook page.

The idea is to transmit colour pictures of Mars and conduct some exo-Mars studies, like attempting some opportunistic imaging of other satellites like Phobos and any comets while moving in the elliptical Martian orbit. This camera has 16 exposure levels out of which there are two or three that are suited for Mars, meaning that it can take other pictures also, like those of some distant planet. It can also provide images of the surface features of Mars with varying resolution and scales using the unique elliptical orbit. This would include craters, mountains, valleys, sedimentary features, volcanic features, rift valleys, and mega faults on the surface of the red planet. Picking up an available commercial lens weighing 640 grams, the SAC engineers disassembled it, took out the lens elements, totally made their own system, and then integrated it, reducing the total weight to 320 gram. Vishnu D Patel, a young engineer, who did the bulk of this work on the MCC, says that 'within a span of four to six months we were able to realise the lens assembly that was used.' Engineers fashioned the MCC bit by bit, in a way that it would collect visual optical light and focus on a detector bought commercially. They also used a specific kind of arrangement of colour filters that were famously put together for digital cameras by Bryce Bayer and globally patented as the Bayer pattern imaging technology, or the Bayer filters. The lens assembly and detector were created so that they ensured compactness, and high speed electronics just near the payload takes care of structural strength despite being light. This is important so that no artefacts are created in the imaging. For all ISRO's remote sensing work, cameras click monochromatic images; in this case the same scene will be shot in three different colours and then intercalated, another new experiment in space technology. Using whatever was available, working in reverse and developing the verification model in a few months, one or two months for proving the concept, Patel believes they met the challenge head on. And the first moments of joy came when the first picture of the Helen storm viewed over India was beamed back, looking good and clear. 'We expect the MCC to give images of the desired quality during the rest of the Mars Mission also,' is what the SAC team feels. That too would need to be controlled, for a variety of reasons. Too many pictures or too much data is as much a problem as no data, otherwise in terms of the capacity of the instrument, there is no limit to how many pictures the MCC can take. It

Standing out on the open horizon is the giant dish antenna of Byalalu, dominating the landscape of the Deccan's flatness.

Credit: Pallava Bagla

can capture an image every second, but data has to be transmitted to the ground and that is what limits the number of pictures that can be taken. In the real environment of deep space, a single picture taken would be about 40 megabytes and it would take at least two to three hours to transfer one image to the ground. This would then need to be corrected and cleaned so that the visualisation is perfect. So selectiveness and being choosy would be critical, and the MCC cannot just keep clicking away. Interestingly, the MCC would also play an important role in providing what is called contextual information for the other science payloads on board the MOM. This would help towards more meaningful interpretation of scientific data from the other sensors. Anyway, the highly dynamic nature of the Martian atmosphere and surface makes the MCC a necessity.[8] Also, as Arya is fond of saying, good and effective outreach of images and pictures will make the tax payer happy, as he or she sees value for their money in such images.

Meanwhile, another major characteristic of the Martian atmosphere is supposed to be the gas, methane, made all the more important because its detection is directly linked to the possibility of life. After being unveiled at a presentation during the hundredth Indian Science Congress in January 2013, the MSM, weighing 2.94 kgs, was finally ready for flight in June 2013. Kurien Mathews, the scientist leading the group getting the sensor ready at SAC, is relieved that they finally managed to design a really high sensitivity optical system that can scan the entire Martian disc within six minutes. The MSM is a first, because the conventional route would have been to use a spectrometer, which would have taken too much time to make. It is what the instrument makers call a 20 bit system with its full signal consisting of nearly 10 lakh counts. This is the first time such an instrument is riding into space, because such a sensor has not been used before. While it saved time, the MSM in its current design has one drawback, that it can only measure methane, and no other gas. Does that make it better for accuracy? Not necessarily, it is only that with limited development resources, the design of the MSM has maximised sensitivity. The SAC team first went about collecting data measurements from the Earth's atmosphere, readying themselves to be able to apply corrections when actual data from Mars would start pouring in.

That is a tantalising possibility. How much methane, if at all, is present in the Martian atmosphere? The methane problem is by far the most important Martian question, with three or four groups having detected methane over Mars, either telescopically or through satellites in the last five to seven years. Over the Earth we know its presence is biological, so if methane is actually present on Mars, is it through a biological source? But it can also be due to comets, therefore emanating from a geological source. The ISRO teams believe that much like their locating water on the Moon through Chandrayaan, if the MOM orbiter detects methane, it would be big. The American Curiosity could not locate methane, although two earlier satellite missions, the Mars Global Surveyor from the US and Mars Express from the EU have detected methane. Both used a spectrometer kind of instrument that cannot distinguish between biologically or geologically originating methane. In fact, Curiosity's roving on Mars in October 2013 led to a somewhat 'tenuous no' on the presence of the gas in that atmosphere, but this too was really not definitive.[9] So, is Martian methane a certainty; is the gas local or global across the planet; what is its source? All these questions are buzzing around the age-old speculation about life on Mars. India is not alone in thinking that there remains value in mounting the methane search. ESA's Trace Gas Orbiter planners are continuing to focus on the elusive gas that has so much hinging on it, and that is a mission due for launch in 2016. Anyway, one possible advantage of the MOM mission is that it is in a highly elliptical orbit, at apogee the satellite is 80,000 kilometre away, so the spatial distribution of methane over Mars can be measured. And since Martian atmosphere is very turbulent, with frequent dust storms, MSM data would need to be collected as often as possible. Any overt emphasis on the link between locating methane and determining whether there is life on the red planet is frowned upon by many purists. Beyond the methane search, scientists are also probing what the surface composition of Mars is like. For RP Singh, who is the Principal Investigator of a project that designed the very special thermometer-like instrument TIS, which at 3.2 kgs is unusually light for instruments of this kind, all this is very exciting. 'By measuring thermal emissions from the surface after processing these signals, we will know what the surface temperature and surface composition is like.' This would help in various ways. One, through detection of hot and cold spots on the planet, which would in turn give a clearer picture of weather conditions on Mars; the other contribution would be its ability to detect minerals, and who would dispute the fact that knowledge about mineral resources is valuable. Going by the notion that there was water on Mars at some point, there is also the possibility of locating minerals that are hydrated. Measuring Martian thermal ambience real-time and thermal emissions from specific minerals and soil types, TIS will assist in piecing together the current picture of the geological surface of Mars, since every mineral has a special thermal signature. Readied in May 2013, this instrument, designed and fabricated completely indigenously at SAC, has infrared optical components that can be difficult to procure particularly in a short time, because they are generally used for defence purposes.

For most people, 'instrumentation development' has a dry, drab sound to it, aligned to the imagery of dreary looking laboratories with all kinds of machines and boxes strewn around. But for Dr Anil Bharadwaj, senior scientist at VSSC and currently Head of the Planetary Sciences branch of the Space Physics Laboratory there, it is a passion. Developing instruments is actually his life. Having been part of the Chandrayaan-1 team, Bhardwaj was then collaborating with

scientists in several countries to develop some of the payloads on the Moon mission. For the MOM, Bhardwaj's team developed a special atom and molecule analyser MENCA a mass spectrometer that is meant to study Mars' upper atmosphere and exosphere, 200 kilometre and above from Mars, something not done earlier. MENCA is designed to pick up atoms and molecules of hydrogen, helium, carbon, nitrogen and oxygen in the Martian atmosphere, and what is more important is that it will also try to unravel the kind of seasonal and diurnal (day and night) variations these gases display. In other words, it will smell and taste upper Martian atmosphere, especially as a function of time. So far, in what we know about the atmosphere of Mars, there are no observations available on the real and live situation with regard to these gases – what are called *in situ* observations. How do these gases vary in concentration as you go away from Mars or come back to the planet? Bhardwaj speaks of how the life of the MOM is also the duration of a season on Mars, because summer, autumn, spring and winter roll over only once in six months, owing to the highly elliptical orbit that the planet travels through.

Why is all of this important? The periodic transitions in the nature of the Universe – especially in the characteristics of Martian environment and atmosphere – are critical to understand. The planet had a very thick atmosphere in the past, so where have all its components gone? Why is the atmosphere so thin now? Where has all the water gone and why is it now so depleted? With Mars being smaller as a planet, and the impact of solar radiation, escape processes are faster, and this rate gets hastened even further by the highly elliptical orbit. CO_2 is dominant and then there are the other gases like nitrogen and oxygen. At what rate are they being lost from the system is another question that scientists are asking, and how are these entities changing in terms of their densities over time? If answers are found, the story of how Martian exosphere is escaping might get pieced together and it may become easier to predict what might happen in the future, which is important if Mars is to be a future destination for humans. In an indirect way, it will also help answer as to why there is so much liquid water on Earth.

The other payload, developed by LEOS, the Laboratory for Electronic and Optical Systems, Bangalore is LAP which is somewhat complementary to MENCA since both the instruments measure atomic hydrogen. Through LAP's measurements, the ratio between heavy and light hydrogen, technically called the Deuterium/Hydrogen ratio may get worked out, and through this a kind of scenario-building is possible, helping scientists discern whether Mars was also like Earth at some earlier time. Developed by ISRO scientist M Viswanatham, LAP will try to build on knowledge about the persistent loss of atmosphere in the Martian environment. The relative abundance of deuterium and hydrogen in the Martian upper atmosphere will help grow human understanding of water loss on the planet. It is believed that the colonisation of Mars, if and when it happens, will be supported by the use of deuterium as fuel with fusion reactors, and this might make the red planet a power-rich economy.[10]

There is a tantalising possibility that emerges from all this science, and that is what excites Goswami the most. He wonders if the next generation might probably think of Mars differently from what has been the current human construct – a remote, red neighbour so far away. It is this power of thinking that gives Goswami the clarity as to why India is on a mission that is sometimes projected by critics as a 'me-too' kind of exercise by the country. This is because he has heard similar critiques when Chandrayaan-1 went up. Who would have thought we would find something new?[11] But, as mentioned earlier, Chandrayaan-1 found water molecules, despite the fact that US and Russian missions had been searching for forty long years! Scientists are also excited that a comet might come near Mars and actually getting to measure the tail of the comet might be a bonus; in fact, this was also a reason to fast-track the mission.[12] Otherwise, special missions have to be mounted to study comets. There is certainly a sense of incredulous wonder at the array of scientific facts that might get unearthed because of these payloads, and nobody doubts that all this information will take forward the human understanding of Mars and its evolution.

Despite all the excitement, ironically, Rao has remained sceptical. Small satellites and small payloads mean that the ideas of exploration are themselves small is his perception, leading him to question how good the experiments really are. The irony is in the fact that he has led ADCOS in its decision-making. All in all, the weight limit was a deciding factor, which as experts like Goswami say, is less than the baggage limit you get when you go on an aeroplane.[13] He is also the same person who has said that 'big' doesn't necessarily mean great science. Either way, there seems to be the usual wave of self-confidence that all the scientific effort is focused on addressing real and burning societal issues, and that the time has come for India to venture into pure space science, pushing the envelope on the call of the frontier. The creation of space-borne instruments that generally address the domains of communication, broadcasting, navigation, disaster monitoring, sensors and

Mars mission launched, two happy space buffs. Pallava Bagla (L) with K Radhakrishnan. After the successful launch of the PSLV on November 5, 2013 the ISRO chief gave only one exclusive interview which was to Pallava Bagla, NDTV's Science Editor.

Credit: Pallava Bagla

transponders among others is the key to national development, particularly for a country like India.[14]

To the scientists and engineers at SAC and VSSC involved in creating the payloads and committed to slogging it out on the floor of the development labs, the esoteric arguments are like distant drum-rolls. Arya likens their work to that of a doctor who has to investigate a patient – 'you refer the patient for a sonography or a cardiogram that would show you signatures of the heart, and you would also go for pathology tests. Similarly, this mission has pictures on the one hand, spectral details on the other, and explores biological or geological possibilities of methane.' Remaining logged in right through, as Sarkar says, is important for them – 'Till at the launch pad, we do a final review, climb up the rocket and ensure everything is in place in terms of the payload.' Then, once the data starts flowing in, its payback time.

The real excitement, doubtless, is in the surge of figures, numbers, graphs and measurements that start to flow once a spacecraft starts the actual work it is meant for. In the case of Mangalyaan, this is all the more exciting, not just because it will be after a really long and strategically planned journey that has its own pitfalls and challenges, but also because of the long gap in communication signals going to and fro. The ISSDC in Byalalu is a nerve centre for the Indian planetary exploration programme because of the significance of data that flows in from satellites as part of missions. This is the place where all data from space mission satellites flows in, is stored using an open archived system of the gold standard in the world of planetary data systems, and is preserved for long-term use while also shared with key stakeholders. This is a science-driven data management approach that can store data for 50 years, can be freely distributed online, and can be made available to all. All components are integrated at I-SITE. It is the first time that ISRO is managing such a long-term mission, with each day characterised by some or other mission health monitoring activity.

Science is only beginning to catch up with science fiction, and the galactic need of capitalising on a window of perfect alignment between the Earth, Mars and the Sun using a sub-optimal low power launcher was a strategic one. After all, the *Chota Bheem* with its big punch glided past its halfway mark in April 2014. Mangalyaan has a lot of hopes riding on it, as the visible result of India's greatest space adventure yet. Till date, only the US, Russia and the European Space Agency have successfully managed to send a spacecraft all the way to Mars. The odds are tough, if the history of Mars missions in the world is the yardstick. There is not a single country in the world that has ever successfully reached Mars on its first attempt. The otherwise highly successful China has also been buffeted by failure. MOM also took some early challenges on the chin, when just a few days into its journey, fuel flow to one of the engines had stopped on November 11, 2013, when the engineers were trying to execute the fourth of five planned, orbit-raising operations. But overall, ever since the MOM took off into space, joy, relief and a sense of satisfaction have been dominant among senior men and women at ISRO, with most on-board systems functioning normally. This leads to a general air of quiet confidence in most of the institutions, and among most people involved, however small or however big may be the task. With his trademark chuckle and in a self-deprecating style Annadurai says, 'The world takes us seriously now, after we found water in the tenuous atmosphere of the Moon.' The time to make a mark on Mars is now, and ISRO seniors often say this is not because India wants to compete with other nations. They see it as a good learning opportunity and their contribution to world space exploration science. All said, in that same world, with its big ideas, big dreams, big machines and infinite distances, the race against time that the Indian Mars mission set out on, seems like an ill-fitting piece. How everybody seemed to have only time on their minds – and a strange obsessiveness about it. All this brought up a critical question – whether this obsession would hit the bull's eye or become the most grievous mistake. But to Annadurai, Arunan and many others, the unambiguous fact is 'we have to serve the country, always'. This has been almost like an under-pinning philosophy of the Indian space programme from the Sarabhai days. Meanwhile, back in the present, what also makes headlines are quaint details, like Radhakrishnan's religious sentiments, and how he visits Tirupati, the iconic south Indian temple town, to pray for the success of the missions he leads. What is incredible is that although he uses a government vehicle to carry the miniature scale models of the rocket and satellite to place at the feet of the idol there, he pays on his own for that part of the trip. Religious belief after all, is a very personal thing. There is also a philosophical twist to how ISRO's leadership views the mission once it is on its way. As Dr Kiran Kumar, Director of SAC, said, 'What you are going to get tomorrow is a big if, and if you think you are not expecting anything, you get something. We are hoping for the best and preparing for the worst.' But Prime Minister Narendra Modi has even greater ambitions. Witnessing a space event in June 2014 at Sriharikota, he said with his rapidly cementing, trademark dramatic oratory: 'We have done a lot but *yeh dil maange more* (the heart wants more). I have proposed to our scientists that they develop a SAARC (South Asian Association for Regional Cooperation) satellite. This will help all our neighbours.' A new challenge for ISRO, one would think!

NOTES

1. A. Lele, 'Mission Mars: India's Quest for the Red Planet', Springer Briefs in Applied Sciences and Technology, 2014.
2. 'Spaceflight 101 – Space News and Beyond', Accessed at http://www.spaceflight101.com/mars-orbiter-mission.html on April 18, 2014.

3. ISRO Mars Orbiter Mission, Executive Summary, Department of Space, Bangalore, Government of India, July 2012.

4. T.S. Subramanian, 'Mars Orbiter Tests have shown our Ability to Predict: ISRO Chairman', *The Hindu*, November 20, 2013.

5. P. Bagla, Interview with K Radhakrishnan, October 18, 2013.

6. Justin Rowlatt, 'Exchanges at the Frontier', BBC World Service, November 5, 2013, with Jitendra Nath Goswami, Director, Physical Research Laboratory.

7. P. Bagla, 'India's Mars Mission is all about Technology', NDTV, Aired on January 4, 2013, published on October 8, 2013, duration: 6 minutes 58 seconds.

8. A.S. Arya et al., 'Mars Color Camera Onboard Mars Orbiter Mission: Scientific Objectives & Earth Imaging Results', 45th Lunar and Planetary Science Conference, 2014.

9. E. Hand, 'Hopes Linger for Mars Methane', *Nature*, vol. 491, November 8, 2012, p. 174.

10. A. Lele, op. cit.

11. Justin Rowlatt, op. cit.

12. P. Bagla, 'Exciting that Mangalyaan May Fly through a Tail of Comet', NDTV, published on October 13, 2013, duration: 7 minutes 43 seconds.

13. Justin Rowlatt, op. cit.

14. A. Lele, op. cit.

Chapter 5

Bigger and Better Rockets

A shop floor on which rockets are fabricated can make human beings feel inconsequential. I actually felt like an ant crawling on the ground. The sheer scale of action gets to you. All around are enormous metallic structures that must have been crafted out of special metals and alloys. I am allowed to step into an unmade rocket, and gazing up through the cavernous column, I am incredulous that this huge and heavy thing, when it is finally ready, will fly out into the skies, effortlessly as a bird. Why fly, it will actually sear its way through the clouds, in a blaze of fire and glory. Considering that human beings can feast their eyes even on an ordinary aeroplane in the sky and derive great pleasure from it, seeing a rocket taking off is a sight that can create immeasurable excitement. But in the picturesque Veli Hills that rise abruptly within the city of Thiruvananthapuram, quite the last mile of the Western Ghats, rocketry is part of a day's work for the engineers and technicians who work at the VSSC, which is revered as the cradle of the Indian space programme. One is struck by the ease with which men in shop floor overalls walk about with purpose, despite the enormity of the stuff that they are handling. Some of them are working on the super-colossal Geosynchronous Satellite Launch Vehicle Mark III (GSLV Mk III), India's flavour-of-the-season rocket that packs in more power and punch than ever before.

For some reason, I get the same larger-than-life sort of feeling walking into the offices of the man who is heading the project for this new and ambitious rocket. The impressive engineer strides across the room, skirting his large table adorned with a miniature rocket with extreme confidence. He shakes hands with a special brand of self-assuredness. As the conversation begins and warms up, I recognise that this is characteristic of S. Somanath, Project Director for the GSLV Mk-III project and Deputy Director of the Structural Engineering Entity at VSSC. He believes in his own self, and in the work he does along with his team. 'I always dreamt of being a rocket engineer, but today my dreams are much bigger than what I am currently achieving.' says Somanath who was one of the key scientists explaining complex facts to an attentive Prime Minister Modi when he visited Sriharikota for the first time in June 2014.

In another part of the same building at VSSC sits another rocket engineer, the difference being that he is the man looking after the trusted Indian rocket, the Polar Satellite Launch Vehicle (PSLV). On another floor sits K Sivan, who looks after the other big rocket programme, the GSLV Mk II. P Kunhikrishnan is unassuming and less flamboyant, but as the Project Director for the PSLV and having been Mission Director for the last eleven successful PSLV missions, is quite literally the man who holds the key at Sriharikota when the final lift-off will happen. 'Personally, I am very happy that the 25th mission of the PSLV was the Mars mission,' he says in a soft voice. Although the PSLV is a trusted vehicle and nothing much needed to be modified for it to be ready for the MOM, except for a new and complex mission sequence and trajectory design. Kunhikrishnan says he and his team did not relax even for a moment because of the spotlight that came with this being the first inter-planetary mission, creating an obvious need for extra caution and stringency. But all this seems to be just a slice-of-life kind of thing at ISRO.

From the plumes of smoke and fire of an Indian rocket, a nascent dream rises. Or is it actually a very old dream? When the British finally left India in 1947, it was as though the clock of an otherwise ancient country had to be reset. Starting all over again wasn't easy, and the struggle to establish self-reliance in several areas of human endeavour one more time, was central to this hardship. But somewhere, there were dreamers who believed it was possible to be self-sustaining and independent, be it food production and agriculture, education, industry or science and technology. That is why this sense of squaring off, when, on June 30, 2014, the PSLV C23, one of the versions of India's most reliable rocket flawlessly hoisted five satellites belonging to some of the world's most advanced nations – France, Canada, Germany and Singapore – 660 kilometre above the Earth. This is a true blue commercial venture. Not so long ago, India used to purchase space images from some of the same countries and lug its satellites abroad for launches. Not so long ago, it would have been unimaginable to think of this country playing host the way it does now to other space-faring nations. Kahlil Gibran's evocative words, 'Trust in dreams, for in them is hidden the gate to eternity' come to mind. A small clutch of highly motivated and passionate people trusted their dreams and those of others, and the gates opened for the Indian space programme, seemingly never ever to close. No wonder Prime Minister Modi said that he felt specially privileged to witness the commercial launch in person, going on to explain why he felt that way. 'This [space] is one domain in which we are at the international cutting edge, a domain in which we have pushed beyond mediocrity to achieve excellence,' he said, adding that this was a 'global endorsement

The old and the new. India's Polar Satellite Launch Vehicle, a marvel of the most modern rocket engineering, with a cyclist in the foreground. Cycles are among the most humble modes of transportation in India and rockets can put satellites into space – such is the magic of life.

Credit: Pallava Bagla

of India's space capability'. Meanwhile, silent but strong on the other launch pad at Sriharikota stands a monster rocket, reaching higher than a 15-storey building, going through its final checks before it readies for flight. This is the GSLV Mk III that, once perfected, can transport really heavy satellites into the Geostationary Transfer Orbit (GTO), which is the pit stop for satellites that have to be eventually moved into the Geostationary Orbit. 'This is our next generation rocket,' says Somanath in a matter-of-fact voice, 'and through this we will increase and double payload capability'. Undeniably, the world's appetite for television, telephones and many other applications of the range of telecommunication services that artificial satellites can provide for humanity seems to grow each day like a giant baby, and if satellites are to keep pace, they need to be larger too. This naturally means the rockets need to be heavier, stronger, bigger and smarter. The PSLV itself has, in its many *avatars* over the years since it first took to space in 1994 with an 800 kgs remote sensing satellite, has expanded its horizon to more than 1,500 kgs as baggage, and gone farther and farther away from Earth.

As outlandish and sci-fi as they may seem, rockets are a modern human necessity. But nothing feels mundane when one witnesses a rocket launch by ISRO. But, rockets are simply vehicles for human welfare, never mind that the technology also has serious defence and military dimensions that are constantly being perfected by the nations of the world in the form of missiles. The much greater reality is that life as people know it today in many, many parts of the world wouldn't be the same if it weren't for rockets lugging satellites into space and leaving them there to beam television programmes, work telephones, predict weather and disasters, among countless other helpful (or should one say indispensable) services of modern life. That is why rocket development at ISRO has seen this almost organic path of progress over some 35 years or so, constantly keeping in mind the fact that new designs would have to keep pace with satellite service needs of India. That is an unquenchable appetite of a 1.2 billion and growing population in a country of enormous geographies and diversities, where often, a satellite can reach where no human can. A 2014 United Nations report[1] recently declared Delhi, India's capital city, as the world's second most populous city with some 25 million people, in a list of world cities where Tokyo is the most populous, and Delhi is followed by Shanghai, Mexico City, Sao Paulo and then Mumbai, at sixth rank with 20 million people! That is why Somanath says 'our requirement for commercial transponders is growing, 500 of them are required across different bands just till 2014–15, for telephone connectivity, Direct-to-Home TV, banking and other digital services, the share market and for dedicated links for our forces, among others.' India's economy and daily life relies heavily on telecommunication satellites and commercial transponders that can be accessed if the country places satellites in the Geostationary Orbit which is the specific orbit for all kinds of communication and meteorological satellites. This very special, high orbit is such that its position remains the same for an observer on Earth who is stationary. A satellite in this orbit seems to stay in the same place because it is orbiting at the same speed at which the Earth is rotating. So, antennae that are built on the ground to communicate with such satellites can remain trained to one spot in the sky rather than moving around to track them. This allows for seamless data flow, with signals beamed up and down without any trouble. 'Slots in the orbit are few,' warns Somanath, explaining that there is 'an international body that decides which slots a country can get and you have to fill the slot in a stipulated time. If you don't fill it you will lose the slot'. Considering the amount of television that Indians are watching now, this is almost becoming an essential service. But the powerful association between a rocket lifting off into the sky and essential services for society needs to be made much more strongly than it is today.

One of the few women whom the author met at ISRO in connection with the Mars mission, N. Valarmathi, who is heading the Digital Systems group at the ISRO Headquarters as Group Director, comes across as being extremely devoted to her work. In her clipped, somewhat text-bookish style she sits down and earnestly explains what motivates her – 'Something unique we must do, so that ordinary people can say that a particular service they are enjoying is actually coming from a spacecraft up in space; that would be great.' Isn't that true? When critics of the Indian space programme and its wasteful expenditure voice their opinion on endless television debates, do they give a moment to think about the extremely sophisticated satellite technology that is allowing their voices to boom across countless living rooms and communities in the country? Rockets and the satellites they carry are the backbone of societal needs in a slice-of-life, essential kind of way, indeed, that is the truth. If it weren't for more and more rockets going up, we wouldn't be able to place spacecrafts out there with their transponders so that all these services with such intense level of applications in everyday modern life could continue uninterrupted. Somanath says: 'If one television channel has a problem, the provider receives so many calls and obviously has to immediately switch to another transponder. Any communication satellite provider faces this problem, so bigger the satellite the better it is. That is why, commercially, it makes sense to have a four to six ton spacecraft, and the economics of shifting to heavier launch vehicles is clear.' Therefore, automatically, the GSLV comes into the picture and so does the need to perfect it as the next generation of rocket technology.

Easier said than done. The GSLV itself, as a concept, is a child born in difficult times, when in the 1990s India was under severe technological sanctions. That is perhaps why it is sometimes like a roguish horse, bolting unannounced and causing some angst to its caretakers off and on, tantalising them into nicknaming it the naughty boy of the ISRO rocket stables, as mentioned earlier. But with the urgency to have rockets that could hoist the 4 ton class of spacecraft into the GTO, in fact, some requirements even go up to 6 and 7 tons, perfecting the GSLV was

something of a non-negotiable for ISRO, and the largely geo-political circumstances created at the time were an obvious catalyst for the Indian programme to get going. Having said that, it took two long decades – which is a lengthy time frame by any standard for organisations like ISRO that move rapidly given their own energy and commitment, and the kind of political and governmental support they get. In the beginning of 2014, in fact in January itself, like a New Year's gift from ISRO's rocket folk, the GSLV Mk II, flying GSAT-14, a communications satellite, managed its first indigenous flight that used an Indian cryogenic engine, and was successful. For K Sivan, who was Mission Director for the GSLV Mk II D5, and for most other rocket scientists at VSSC, the January 2014 mission brought with it multiple successes.

प्रथम प्रमोचन मंच
FIRST LAUNCH PAD

1 to 4: Views of the Polar Satellite Launch Vehicle on its silver jubilee mission to carry India's maiden mission to Mars. The mission was launched from the First Launch Pad at Sriharikota.

Credit: Pallava Bagla

Back in August, 2013, the same rocket was ready for launch when a leak was observed in the second stage of the GSLV Mk II. Scientists discovered that 750 kgs of highly inflammable and explosive fuel had leaked from the engine, and the countdown clock was stopped barely 74 minutes before the scheduled lift-off. Weighing as much as 80 well-fed adult elephants, the vehicle was dismantled at Sriharikota. Chairman Radhakrishnan said, 'timely detection and quick action' averted what could have been a massive fire that could have engulfed not just the rocket, but the second launch pad facility at Sriharikota as well. This is the first time ISRO has had to dismantle a GSLV rocket that had been almost fully fuelled. Back then, the Chairman had also added the assurance that 'there is no generic problem with the rocket', and that the launch would be attempted again in December.[2] The January 2014 launch was also special because it was the first rocket carrying India's domestically designed and produced cryogenic engine for the first time after 20 years of working on it. Two decades of effort and ISRO developed the cryogenic engine and had it flying on the GSLV Mk II. The workability of the indigenously-developed cryogenic engine was obviously a very significant milestone. It is, after all, this technology that was denied to India 20 years ago, making it not just among the most written-about and highlighted sagas of India's space programme, but also a dramatic story of development on its own, a story that has been central to the path that rocket development has taken at VSSC in the last few decades.

In the southern spur of the Western Ghats is a hilly place called Mahendragiri, in Tamil Nadu. Here, set in emerald green surroundings is the ISRO Propulsion Complex (IPRC) that now assembles, integrates, and tests ISRO's cryogenic engines. With the Indian government having provided close to ₹25,000 million for the development of the GSLV Mk III, a sizeable share of which (₹15,000 million) went into facilities and infrastructure development, and the cryogenic facility was among the ones that came up. In its development, the cryogenic engine and its attendant facilities tell a dramatic story. The science of producing ultra-cold temperatures and studying what happens to stuff at those temperatures is cryogenics and cryogenic engines are much more efficient for use in rockets than solid or liquid propellants. Complex and hard to crack, cryogenic engine technology raises further hackles in India owing to certain troubling historical events of the 1990s. On the verge of receiving engines and a subsequent transfer of technology from Russia's Glavkosmos space agency, India was suddenly faced with cancellation of the contract in 1991, through a *force majeure* clause. Missile Technology Control Regime Guidelines were cited, and the US's involvement was also noted.[3] Later in that decade, even greater exacting sanctions were laid on India, after the Pokharan blasts. Simultaneously, the urgency to design and create launch vehicles that could carry heavy satellites into the GTO were also building up. Naturally, efforts to indigenously develop the cryogenic engine and its technology began. Liquid hydrogen and liquid oxygen are the most common cryogenic propellants, and a cryogenic engine is known to be extremely powerful in carrying heavy baggage into outer space. This was obviously the key to developing the GSLV series of rockets and finally, when the GSLV Mk II managed a smooth lift-off, ISRO's relief was understandable. There were early versions of the GSLV that flew successfully by using Russian cryogenic engines, but since technology transfer had been denied, ISRO always knew that it was on borrowed time and needed to crack the technology, rather than continue to rely on ready-made imports, which were anyway for a limited period.

It was around 2000–2002 when led by S Ramakrishnan, Director, VSSC, Somanath and others started asking themselves the question as to what new design could be brought in so that they could push launch capacity, which was then at a maximum of two tons, to at least another two tons. They knew that it could not just be a variant of existing machines, or an upgrade, and that is how the GSLV Mk III got conceived, a novel piece of engineering that acknowledged the design concepts behind the versatility and reliability of the PSLV. Now, after many years and some trials and some disappointments, Somanath and his team wait for the day when the monster will take off into space. 'The two ton class of satellites need double the power of what the PSLV has, and this kind of energy content for the vehicle comes from the cryogenic engine,' says Sivan who anyway believes that a launch vehicle has to be perfected as a vehicle, not for a specific project. With satellite size having grown in recent years, the GSLV Mk III is an essential member of the launcher family. The GSLV Mk II and III are in different categories, that is why both are important. 'It is better to have all the vehicles as available options, instead of wasting cost using bigger vehicles for smaller satellites,' Sivan explains. So, neither is there redundancy in the GSLV Mk II and nor does it in anyway keep from the GSLV Mk III being flight ready. This basket of choice allows ISRO the freedom to select a vehicle aligned to a specific demand. As Sivan says, 'Our technology and hardware is available; so whatever is required, we can go ahead.' While technology denial and isolation played a huge role in the protracted developmental history of this rocket, even in terms of concept and design, the technology is hardly simple, involving three kinds of engines, with solid, liquid and cryogenic propellants.[4] With perceptible passion, Sivan said late in 2013: 'We have data from seven flights, and have analysed each and every bit, validated and taken corrective measures with high altitude tests. Once on track this rocket (the GSLV Mk II) will become a highly reliable and simple vehicle. For now, we need to remove the bad name, and we will have a great quantum jump in the space arena.' Somanath feels the same way – 'The two ton class of spacecraft are slowly fading away, and with the PSLV as the only mainstay vehicle it is very important to master the GSLV Mk III.' And so it stands, ready and waiting, to carry into space a dummy spacecraft that happens to be the newly designed and tested crew module that ISRO has readied in anticipation of the Human Spaceflight Programme (HSP)

Surreal landscapes, an observatory and a Buddhist monk with his prayer wheel. Hanle in Ladakh is home to one of the world's highest astronomical observatories of recent times. Located at an altitude of 4,500 metres in the cold desert of Ladakh, Hanle houses the Himalayan Chandra telescope, a 2.01 metre optical masterpiece which makes use of the rarefied atmosphere and dark nights to study the Universe. This observatory is managed by the Indian Institute of Astrophysics in Bangalore and can be remotely controlled using satellite connectivity.

Credit: Pallava Bagla

Rocket of the future. The Geosynchronous Satellite Vehicle Mk III, in a mock-up exercise on the Second Launch Pad at Sriharikota. An experimental launch puts this monster rocket to test.

Credit: ISRO

India's first Satellite Launch Vehicle 3 (SLV-3). This 22 metre high, 17 ton vehicle was a four stage rocket capable of placing 40 kg weight satellites into space. This vehicle had four flights of which two were successful and two were failures. The first flight took place on August 10, 1979, from Sriharikota. The SLV-3 project was headed by India's former President APJ Abdul Kalam, who then worked at ISRO as an aeronautics engineer.

Credit: ISRO

taking off in the near future. The new rocket has just four propulsion modules compared to the seven in the GSLV Mk II, with each far more powerful than their earlier avatars.

At this momentous time in the history of rocketry in India, when engineering advancements are moving rapidly, there seems to be something new on offer all the time from the celebrated rocket factories in the lovely Veli Hills of Kerala. From the pencil-like early rockets to the Goliaths of today, rockets have been growing, changing, becoming more refined and increasingly efficient in their launch capability. This surely makes a lot of commercial sense, with India rapidly gaining ground as a preferred destination in the commercial, multi-million dollar launch market of the world. In a recent tally, the PSLV alone has launched 40 foreign satellites from 19 countries, including today's developed nations. But what is really intriguing is the kind of poetic justice there is in the dramatic path rocketry has taken in India. President APJ Abdul Kalam, perhaps the first-ever formal Project Director of the rocket-making activities at ISRO, has often noted how strange it felt for him to see a lovely painting of Tipu Sultan's army fighting colonial British troops mounted prominently at the reception of the US's Wallops Flight Facility situated on Wallops Island, off Virginia's eastern shore. This is a famous painting that shows soldiers launching rockets. Tipu Sultan, the Tiger of Mysore as he was known, and his father, the equally illustrious Haider

Ali, were considered pioneers of rocket development in India, more than 200 years ago. Professor Roddam Narasimha, an aerospace scientist who is now with the National Institute of Advanced Studies (NIAS), in Bangalore, has studied extensively the early work of Tipu Sultan regarding rocketry. He often says that it is rockets that helped the Mysore army achieve victory over the British in 1780.[5] The glory of this battle is depicted in a mural that is still to be found at the Srirangapatna summer palace in Karnataka. Kalam, of course, was then struck by this glorification of an old Indian practice and how such historical detail was being symbolised far more powerfully on American soil than back home. Kalam has shared this feeling in his autobiography, *Wings of Fire*, saying it made him happy to see an Indian glorified by NASA as a hero of warfare rocketry. Coming from one of India's foremost rocket scientists, this statement is loaded with meaning. President Kalam was assigned the leadership of India's first home-grown rocket-making activity, the Satellite Launch Vehicle (SLV) Programme. After having broken a psychological barrier with the early TERLS launches where, even though they borrowed rockets, India experienced what it was like to plan and execute a rocket launch;

The Augmented Satellite Launch Vehicle or ASLV that had its first flight on March 24, 1987 from Sriharikota. This was a five-stage rocket, developed by ISRO to place 150 kg satellites into orbit. This vehicle has since been retired by ISRO.

Credit: ISRO

A magnificent view of the First Launch Pad before the Mars mission was launched.
Credit: ISRO

that is when India felt ready to get into developing indigenous satellite launch capability for communications, remote sensing and meteorology. The SLV-3 was India's first experimental satellite launch vehicle, an 'all solid, four stage vehicle weighing 17 ton with a height of 22 metre and capable of placing 40 kgs class payloads in Low Earth Orbit (LEO)'.[6] The first flight was a failure, in August 1979. Then on July 18, 1980 the Rohini satellite, RS-1, was placed in orbit from Sriharikota. This put India into the space-farers club, as the sixth country. Two more launches ensued, in May 1981 and April 1983, placing Rohini satellites carrying remote sensing sensors into orbit. With the SLV-3 and its success, the stage got set for the Augmented Satellite Launch Vehicle (ASLV), followed by the PSLV and the GSLV.[7] Kalam, for obvious reasons, came to be closely associated with Indian rocketry, and later transitioned, quite naturally, into India's missile man through his many years of work at the Defence Research and Development Organisation (DRDO). The Rocket Corps of Tipu Sultan's army were about 5,000 strong, according to Narasimha, who also describes how Tipu was a technology buff and promoted the manufacture of rockets in parts of the town that he called *Tara Mandalpet,* that could be translated loosely to mean a galaxy bazaar![8]

Kalam's is a story of rising from the humblest of backgrounds, of which he remains extremely vocal and proud even at the age of 82; it is also a story that is central to rocketry in India. In a poignant description of his two lives and how deeply intertwined they were, he questions as to what was more momentous in his life – the ability to craft rockets, or the boats that his father sailed in the Bay of Bengal. He writes in his recent book, *My Journey: Transforming Dreams into Actions*: 'Watching the boat come to life from pieces of wood and metal was perhaps my first introduction to the world of engineering. Wood was procured and Ahmed Jalalluddin, a cousin, arrived to help my father out. Every day, I would wait impatiently till I could go to the place where the boat was taking shape. Long pieces of wood were cut into the required shape, dried, smoothened and then joined together. Wood-fires seasoned the wood that made up the hull and the bulkheads. Slowly the bottom, then the sides and the hull began to form in front of our eyes. Many years later, in my work, I would learn how to make rockets and missiles.' Kalam refers to how momentous a role boat-engineering played in their lives. 'When I struggled to give shape to the satellite launch vehicle (SLV) rocket, or the Prithvi and Agni missiles, when countdowns and take-offs were disrupted, and rain came down on our rocket launch sites situated by the sea in Thumba and Chandipur, I always remembered the look on my father's face the day after the storm. It was an acknowledgement of the power of nature, of knowing what it means to live by the sea and make your living from it. Of knowing that there is a larger energy and force that can crush our ambitions and plans in the blink of an eye, and that the only way to survive is to face your troubles and rebuild your life.'

Just like a rugged boat that stands its ground against the forces of the unpredictable sea, the PSLV, commonly referred to as ISRO's trusted work-horse, has brought stability, reliability, and a rock-steady quality to India's launch capacity, and this rocket has seen ISRO through innumerable launches, with just the first being a failure. That is why space programme leaders are fond of saying that the PSLV has given them 'successively successful launches'. So it was, when the Mars mission was announced, that the organisation decided to prioritise the mission itself, despite knowing that the GSLV would certainly have been the preferred choice for a long, inter-planetary journey. But it was far from flight-ready. Sivan describes the moment: 'For the Mars mission we first thought only the GSLV could be used. Then we thought, is it not possible to conduct the mission using the PSLV? Then we devised a new and novel strategy of getting into orbit. We understood that the PSLV could do the job with this novel strategy, and decided to go ahead. It was better than waiting for the GSLV. Anyway, any mission we undertake for the first time, the first vehicle will be an experimental one, and that data would be very valuable.'

Finally, it was decided that the PSLV C 25 in its XL avatar – the most powerful version of the rocket – would be the vehicle of choice for the MOM. Primarily developed to launch remote sensing satellites of the 1,200 kgs class into the Sun-Synchronous Polar Orbit (SSPO), this versatile machine can also carry out LEO and GTO missions. It is fascinating how most of the engineers I meet at the VSSC are people who joined the PSLV and its many projects as a first assignment, particularly the currently active leadership cadre who may have joined

P. Kunhikrishnan, Mission Director of the PSLV. Minutes before the launch, the Mission Director has to turn the Mission Authorisation Key to initiate the ignition sequence that triggers the final launch.

Credit: Pallava Bagla

the organisation. Kunhikrishnan, PSLV's Mission Director, sounds like a proud father when he says, 'If you see the PSLV, it was initially developed to launch satellites into polar orbit, then we started utilising its versatility by changing mission parameters. In fact every mission of the PSLV is unique with some type of new requirement of the spacecraft or the user community that has to be catered to.' But surely, considering the enormous length and arduousness of the Martian journey, wouldn't the GSLV have been a better choice and has the decision to use the PSLV compromised the mission in any way? Somanath is clear in his mind about the options, the decision exercised, and its final impact – 'Of course a heavier lift vehicle would have been a better choice, propellant budgeting would have been more efficient, we would have

India's Geosynchronous Satellite Launch Vehicle (GSLV) which is 49 metres tall and weighs 402 ton and can launch communications satellites weighing up to 2.5 ton. Seen here on the Second Launch Pad at Sriharikota. This vehicle has seen more failures than successes for ISRO.

Credit: ISRO

had additional mass for payloads, so we would have got added flexibility. But the mission was of prime importance, and today we have only the PSLV. What we are trying to prove is our ability to design and put a space craft around Mars, that is 95 per cent of the mission objective, the scientific purpose is an add-on. PSLV was the fastest route, if we had waited, we would also have waited to perfect the technique [of executing an inter-planetary mission]. PSLV is half the cost of the GSLV and I am sure that nothing great is lost. These are anyway managerial decisions and it also brought out a novel way of doing it.' Was the PSLV C 25 that carried the Mars Orbiter special? 'For us each mission is new and special, we cannot ever relax,' is what Kunhikrishnan feels. This is despite the strong design of the vehicle and its exemplary track record. 'But this cannot alone sustain success,' says Kunhikrishnan, emphasising that 'the analysis of data for any non-conformity, with even the most minor observation analysed threadbare, is critical, and one of the main factors for sustaining success is the powerful review process we follow'.

Generally, reviews at ISRO are very strong, and have often been cited as one of the major reasons for the spectacular growth and success of the public sector organisation. Also, when adapting a vehicle for a new mission, the general *mantra* is to not change the working system to the extent possible. But Kunhikrishnan knows that this is not to be followed as an unbending rule. 'For example, with the MOM, orbit requirements from the Mars satellite team were very unique and special and had to be delivered by the PSLV. We had to change the total mission sequence and it called for a long coasting period of 28 minutes, so the design was appropriately modified to satisfy this condition', says Kunhikrishnan. Several technical issues had to be kept in mind with respect to this period of long coasting, that the rocket would be in a shadow region for a long time, so temperatures would be really low, making thermal control for the upper stages important. This meant a lot of changes. Then, there was the stringent schedule requirement – fifteen months in total – and the launch had to take place in a very short window. He was also anxious about the timing of the launch, since cyclones tend to hit the coastal regions around Sriharikota during the months of October-November. 'We were doing a November launch for the first time. After October, we normally never plan launches at Sriharikota, owing to weather reasons.' This was indeed a difficult and challenging time for his team. 'Overall, we felt we were under the spotlight because this was an inter-planetary mission, so some things were different, not that we do not have the maturity to handle the pressure!' is Kunhikrishnan's razor-sharp summary of what it must have been like to work on the rocket to ready it for the Mars mission. 'Taking up this mission was a challenge, we had never before executed a mission of this kind with any special orbital requirement; moreover, given the mission's importance we took a lot of care, working to avoid even minor deviations that might have been acceptable.' Meanwhile, other work continues for the workhorse. The Indian Regional Navigational Satellite System

spacecraft were prioritised for launch because 'we want to complete the constellation with seven satellites,' says Kunhikrishnan. The gentle-mannered man is keen to emphasise that 'there is never any complacency with the PSLV, each time it flies. Its success record is so impressive, that is a reason for the pressure.' Indeed. Extreme pressure was to mount when PM Modi decided to witness the commercial launch of the 714 kgs French Earth observation satellite SPOT-7,[9] along with four others. The PSLV C 23 lifted off at 9:52 a.m. and its flawlessness once again vindicated what Kunhikrishnan experiences every day at work, and that the PSLV pedigree is invaluable. Here is a man who has directly handled nine missions of the PSLV onwards of the C 15. The rocket has variants, beginning with the generic type of PSLV that has one core and six strap-ons, and ISRO has conducted 11 missions using this generic version. The PSLV XL version, enhanced in such a way that it could carry payloads up to 1,700 kgs, was used for Chandrayaan-1. The PSLV, in its varied *avatars*, has already launched more than 70 satellites, of which more than 50 per cent are satellites of other nations. The versatility of the PSLV is its strength, as is its fundamental design (see Annexure 6). In fact, the design and its development and testing, all of which is indigenous, are the backbone of this popular well-performing rocket.

This growing wave of indigenisation and self-reliance means a lot for a country like India. Koshy M George, who is in charge of Materials and Mechanical Engineering at VSSC, has met people from a small company in Hyderabad that does some forging work for ISRO's rockets and they tell him how thrilled they feel when they see a rocket going up. 'Every flight gives the same thrill to us since we do the hardware and machinery.' These are the heart-warming trajectories that are getting charted within the rocket-making realms of India's space programme, and the predominant themes of these paths are the almost full scale of indigenisation, and a growing, healthy industry interface. In the conversations with several ISRO experts and senior staff when they speak of the ever-growing trend of indigenisation, it is clear that global equations are shifting rapidly. 'Indigenisation has been progressing very well and dependency has been shrinking. The idea is to have total self-reliance,' is Kunhikrishnan's analysis. He is fond of giving the example of the flight computers on the PSLV, the Vikram range, also developed by his team. Koshy, who has watched this trend very carefully, says that 'through the last decade, we have worked on indigenisation and our hardware is largely indigenous. Not so long ago, in 2004, almost 90 per cent materials were being imported, aluminium and super alloys, thousands of things like fasteners and the like. Today, our imports are just about 10 per cent, 90 per cent materials are indigenous.' This has been possible because Indian industry has also risen to the occasion. Rockets need a lot of special materials and mechanical engineering approaches, and most of these are being produced indigenously. Maraging steel, for example, is a special kind of steel far superior to ordinary steel and is particularly suited for aeronautical engineering. With low carbon content, this steel has nickel, titanium and a few other elements; its chemical composition, preparation and refinement are also special. It costs ₹5,000 per kg while normal steel is ₹50 per kg. Koshy remembers how indigenous production of this special steel was slowly fostered and strengthened, with ISRO working closely with a large public sector enterprise called the Mishra Dhatu Nigam or Midhani, in Hyderabad. 'Once they stream-lined production, many foreign producers wanted to supply us with this steel, but we had already indigenised and were happy with the Midhani work.' The Midhani website has a list of partners on its website and ISRO tops that. Another special material called Titanium Sponge was also not available in India, but is now manufactured through collaboration between the Indian government and the Kerala state government.

The ISRO-industry partnership, over time, has come to be considered as something of a model because it has created a win-win situation for both. The industry is enabled with generous investments to develop newer and newer products, while freeing ISRO's engineering and technical teams from some of the more monotonous and assembly-line type of work, allowing them to focus on the more cutting-edge and creative design-related efforts that are always required. For Kunhikrishnan, there are goals here. 'Finally, we want to have PSLV stages realised by industry and sent straight to Sriharikota, but that is the ultimate.' It goes without saying that quality assurance needs to be retained and that is perhaps why it will always remain more of a partnership than a complete out-sourcing model of working. 'Year after year, we are increasing the participation of industry. Our Chairman has been taking a lot of interest to see that PSLV finally comes out of Indian industry. Even now, 175 industries are participating in the process, huge companies and even smaller ones. They are able to meet our space quality requirements. I would say we are working very well with industry, and we are getting good support. It is a good experience. The costs may be slightly more but there will be eventual gains', is Kunhikrishnan's take on the industry interface. Doubtless, there is a human resource angle to the gain, and manpower constraints are making the industry tie-ups almost mandatory for ISRO. Staff intake is apparently insufficient for ISRO; in fact, Somanath explains that the actual number of engineers has remained roughly the same, at about 3,500 people, since the time he joined the organisation in the mid-1980s. 'Youngsters are coming in but only to fill slots freed up by those who retire, and there is a government ban on recruitment. This is when additional projects are being taken up, and there are roughly four launches per year. That is why we have gone to contractors,' is Somanath's viewpoint. Kunhikrishnan says, 'they deliver systems and our work on the PSLV has come down drastically. We happily hand over, so we can do the more creative work, not just routine work. The industry has slowly moved from manufacturing components, to stage supplies, and the full vehicle is not far.' And so, the tantalising question that he asks is: 'Why don't we sub-contract the entire PSLV? It is fully standardised and fully ready – we can send it out. At least the main vehicle can come straight from industry to Sriharikota in another few years.' MV Kotwal, Board Member, Larsen & Toubro and President, Heavy Engineering says, 'If ISRO is interested in partnering the Indian Industry in overall management of the entire Polar Satellite Launch Vehicle programme and the communications satellite fabrication programme, it is a very welcome and necessary step. With its wide experience and expertise in project management, L&T would be positively inclined and well placed to take over complete programme management with technical support from ISRO. In future we could jointly explore global opportunities in these domains.' The variety of industry partnerships that ISRO has ranges from the

1 to 9: ISRO's naughty boy, finally tamed. On January 5, 2014, a fully indigenously developed GSLV including an Indian-made cryogenic engine had a successful lift-off from Sriharikota.

Credit: ISRO

public and private sector, large and small companies, and diverse geographies across India. And it is a growing family. As Radhakrishnan, known to lead from the front on industry collaboration, says, 'Capability is there, demand is there, now how do we enhance the capacity to realise more PSLV's? As of now more than 400 industrial firms are working for meeting all of ISRO's requirements. Can we get the Indian space industry to realise the PSLV vehicle itself?'[10] Anyway, it is obvious that the way forward would be a consortium of industries that takes charge, for example, of delivering the PSLV, launch-ready, at Sriharikota. But Somanath remains cautious, saying that 'now industry has changed and is focused on quality, but there is still a long way to go.'

In a few decades, for a nation to go from nothing except the romantic notion of the rockets of the Tiger of Mysore, to being acknowledged in the multi-million dollar world of commercial rocket launches, is something remarkable. Kunhikrishnan remembers the time when the PSLV carried 10 satellites at one point, considered a world record of dropping so many spacecraft into orbit. In 2013, there was a dedicated launch for the French. The French are considered leaders in the field of remote sensing satellites, and ISRO launched Spot 6 for them. 'Based on the performance, they gave ISRO another order, for launching Spot 7,' says Kunhikrishnan. This of course, was executed in style. From the perspective of other nations, India's rockets are reliable, far more accessible than the launch services of the US for example, and the Indian space programme provides services in all respects, taking good care and interest. The rocket-launching teams also give 'equal importance to supporting universities and student communities in general to build small satellites and gather their own experience. Stud-sat, IIT Kanpur's *Jugnu*, one from SRM University, there are many, and five university satellites are waiting for the PSLV. These student satellites are launched in collaborative fashion, there is no money involved,' Kunhikrishnan explains happily. PSLV has very high potential for commercial operation and this has been proved time and again. 'We have tagged missions till PSLV C 50,' he adds, saying that a good number of those coming up are for India only. There is a new project to enhance the number of vehicles, and government sanction is available up till the PSLV C 50. Launch frequency has obviously grown, and Kunhikrishnan is careful with his team that is always under vigilance, with every technical area having a designated Deputy Project Director, several project managers and engineers. This is a multi-centre, multi-disciplinary activity, and his core team has to interact with several system development agencies.

Who can doubt that the rockets, and therefore VSSC, are very special for the Indian space programme. If it wasn't for the early start that Sarabhai and others gave to rocketry in India, one dares say that the programme would have been in a very different place today. In fact,

Sarabhai himself is known to have said – 'The sounding rocket programme served as the nucleus of a new culture where a large group of persons in diverse activities learn to work together for the accomplishment of a single objective'.[11] John P Zachariah, Director, Research and Development at VSSC, who has completed almost four decades with the institution, has a crystal clear memory of the early years, when, slowly recognising the need for capability to work on propulsion, avionics, pyro-systems, compressions, stage auxiliary systems and so many other realms of rocketry, infrastructure got built step by step. 'Earlier, we had launched Aryabhata and other satellites, but we went abroad for launching capacity. I was indeed fortunate to be in that first SLV-3 team with Kalam who was a good leader and I was a young bachelor.' Many VSSC old hands like Zachariah gave it their all, some of them putting personal lives on hold. Kalam was quite the task master, and the first launch was not a success. The story goes that Professor Satish Dhawan and Dr Brahm Prakash, who was the

The Space Capsule from the recovery experiment was clothed in a special jacket of silica tiles, indigenously made. These specialised tiles can withstand atmospheric re-entry temperatures that can go up to about 4,000–5,000° Centigrade while the inner temperature is maintained below 50° Centigrade. Pallava Bagla tests one such tile which has been heated on one side with a welding torch, while the other side is still cool enough to be touched.

Credit: Pallava Bagla

first Director of VSSC and the man credited with initiating work in India to build and launch its own satellites, buoyed Kalam's spirits, advising him not to be disheartened. After this came the Augmented Satellite Launch Vehicle series, and then the PSLV, with Dr S Srinivasan as Project Director. Zachariah recalls how everything went well on the first flight, except for a small error in the computer system. 'It was a minor error, but gave us a lot of lessons, how to validate this complex software and use simulation. That gives real strength in avionics, and since then till date we have not had a single failure due to computer errors. We have developed very rugged systems.'

Considering the long history of rocketry, it has taken quite long for India to reach self-reliance, but then that is because technology development for rockets is always cast over by the tricky duality of its purpose – both civil and military. So, generosity is not a feature of nations when it comes to sharing such technology, and this further drives them to indigenise and create fresh technology. Every rocket that takes flight has countless parts to it, and something in the way the Indian space programme is positioned gives to each person who may have had some role to play in constructing the flying machine a great sense of achievement while watching that flight. It must be said in praise of ISRO that it has mastered the art of making people feel good about their work, whatever the scale, whatever the contribution. Many of the senior-most staff at VSSC describe the magical, motivating capacity that was there in Kalam's leadership, and that it has played a big role in the progress of India's rocketry programme. Sarabhai, who was a life force in Kalam's life, would have been greatly pleased with the fact that the Indian space programme is today working the commercial launch market, mastering state-of-the-art rocketry, and visualising a day in the not-so-distant future when the favourite PSLV will be completely manufactured and assembled by an industrial consortium, delivered at ISRO's door at the Sriharikota launch pads, ready to lug some satellite or the other into space. Indeed, these are exciting times, 200 years after Tipu Sultan and Haider Ali's ingenuity, and half a century after the bricks-and-mortar work of Sarabhai, Dhawan and Bhram Prakash. Celebrating the fiftieth anniversary of TERLS in 2013 with lots of interesting outreach activities involving young people, it seems befitting for VSSC to be focusing on newer challenges. The HSP, although not on the immediate horizon if you look for formal clearances, would need what is a man-rated vehicle, and no deviations would be accepted. Sample tests will be many, many more than for other vehicles, and yet, materials and design would be of set standards. In the VSSC of the twenty-first century, Somanath's clarion call is the best reflection of their mood – 'Give us the challenge, and the purpose for design, we can do something new; of course budgets, manpower and political support are required.' If Prime Minister Modi's early positioning is to be taken seriously, and given the history of the kind of government and political support the space programme has managed to garner right through its development, the new challenges will keep coming, and the ISRO rockets will continue to blaze their exhilarating trail through the skies.

NOTES

1. World Urbanisation Prospects, United Nations, 2014.

2. P. Bagla, August 27, 2013. www.ndtv.com. What Stopped India's Rocket Launch? 750 Kg of Leaking Fuel.

3. T.S. Subramanian, 'Cryogenic Success', *Frontline*, February 2, 2014.

4. A. Lele, 'GSLV-D5 Success: A Major "Booster" to India's Space Program', January 6, 2014, http://www.thespacereview.com/article/2428/1.

5. R. Narasimha, 'Rocketing from the Galaxy Bazar. Millennium Essay', *Nature*, vol. 400, July 8, 1999.

6. http://www.isro.org/launchvehicles/launchvehicles.aspx#SLV3. Accessed on July 8, 2014.

7. http://isrohq.vssc.gov.in

8. R. Narasimha, op. cit.

9. P. Bagla, June 30, 2014. www.ndtv.com.

10. P. Bagla, ISRO's PSLV up for sale to Indian Industry. June 30, 2012. www.ndtv.com.

11. P.V.M. Rao and P. Radhakrishnan, *A Brief History of Rocketry in ISRO*, University Press, 2012.

Chapter 6

Getting It Done:
The Men and Women
Behind the Project

Bangalore, Thiruvananthapuram, and Ahmedabad. Three cities that are studies in contrast, which is not surprising, knowing the melting pot that is India. Despite the contrasts, they share an inextricable link, each being an integral part of the Indian space programme. Thiruvananthapuram, ISRO's rocket city and quite the birthplace of the Indian programme is obviously special because it is Kerala's capital city, pretty much the southern-most point in India and a location that Sarabhai chose with care. Ahmedabad, the first city in India to give a home to the Indian space programme, is exciting, irrepressible and intriguing, with a unique mix of congeniality, colour and culture. I assign this attraction to my great and unexplained connect with all things Gujarati – from the riot of colour on their fabric and embroidery, the fabulous 'farsan' that seems created just for my palate, flavours that explode on my tongue, to my great respect for the Gujarati's adventurism. And of course, my Bangalore, with its quaintness and gabled houses that still peep out here and there, the jasmine and *kanakambaram* flowers, the massive, spreading gulmohars, and the now somewhat messy roads where traffic seems to burst out of everywhere, threatening to overwhelm an otherwise incredibly pleasant city.

There are times in life when some of these associations get altered, and the travels that led to this book managed to bring that change, in a small yet potent way. Men and women, some young, some not so young, who have grown out of simple backgrounds and cultures, but are now launching rockets for India, crafting space machines, creating

prototypes for space suits that Indian men and women would wear when the country gets ready for sending humans into space, or designing small equipment to chase big scientific questions. They are the unlikely, even unsuspecting, heroes of a movement called the Indian space programme. And that is why heroism sits uncomfortably on their shoulders, something they prefer to shrug away rather than bask in. After the Armed Forces and civil society movements, this is the only work environment where one observes extreme, even inexplicable, devotion to the job, driven by almost unseen motivation. How one wishes the same drive and determination could become a hallmark of other systems and institutions too, a desire driven by close encounters and experiences with the Indian public health system. The government-owned health sector is expected to provide high quality, accessible, affordable healthcare to all Indians regardless of their socio-economic status There is a grave and significant shortfall here that is primarily because of inefficiencies and mismanagement of the system. It is therefore most natural for a person who has worked for years within the health sector to have a golden wish for its future – that the public health system will be run with the same drive, motivation and efficiency that the space programme is in India. One dares say that would be the day that India would truly be able to call itself a global super power where a super-efficient health system competes in its efficiency with the operational perfection of the squeaky clean ISRO. After all, they are connected as two systems that offer the promise of a better life for the people of India. Moreover, ISRO contributes to the health sector by providing access to telemedicine services. The ISRO folks I meet – most of them – have never even considered leaving their institution for any other workplace, landing up as mere 22–23 year olds for their first job ever after graduating from a neighbourhood engineering college as the first-generation engineers of their families, in the small towns where they grow up. A surge of affection washes over me as I spend time with these people because simplicity is almost like a proud badge that everybody is very happy to wear here. Here it seems the spotlight has deliberately been trained away from the individual despite the fact that exemplars like Sarabhai live on as legends. The focus is on teamwork, and nothing in the realm of space exploration can be achieved without working in teams, often very large groups of people spread over different cities, labs and offices. But somewhere in the maze of the ISRO collective, are some very special people who have run the Mars marathon, directly or indirectly.

For a programme that relies so heavily on machines of all shapes and sizes – from rockets to satellites to payloads – ISRO is unflagging in its people-centred philosophy. That it is the men and women who matter, and the rest is just machinery, which will anyway do its job, is evident. From new recruits to those about to retire, each person is made to feel wanted,

Team ISRO with an almost ready Mars Orbiter satellite, inside the clean room in Bangalore.
Credit: Pallava Bagla

useful, and productive. Interestingly, recent years have seen a surge in recruitments, especially after the success of the Moon mission, Chandrayaan-1. Seniors at ISRO claim that even IIT graduates, who have earlier been known to shun ISRO jobs, have made a beeline to work with the government space programme and ISRO.

When the people of an organisation as large and spread out as ISRO seem like they are all following the same silent tune, it is most likely kudos to the induction-level orientation they might be going through. Otherwise, it is quite hard to think up reasons why a scientist who has spent some 10 years at SAC in Ahmedabad would have exactly the same feelings and motivation for the job he or she does, as does a rocket engineer nearing his retirement in faraway Thiruvananthapuram's VSSC. There is certainly a contrarian view on this pattern, that such sameness of thought could be because of a monotonous lack of originality, creativity or passion. But, it is perhaps more to do with ISRO's people management principles where everybody knows his or her place and role in the organisation, feels good about it, and understands his or her contribution to making satellites, launching rockets, or designing

Leadership at ISRO follows a distinctive culture of high levels of delegation and the encouragement of team work. The stars of ISRO, Chairmen, past and present. L to R: UR Rao, Madhavan Nair, Kasturirangan and Radhakrishnan.
Credit: ISRO

scientific experiments and payloads. What's more, the organisation understands it too, and acknowledgement is generous, rapid and precise. When Professor Dhawan was Chairman, ISRO, Abdul Kalam and his team had worked on the SLV-3 project and the launch was a failure. Following the basic philosophy of good leadership, Dhawan took the mike and responded to all the questions, with his trademark clarity and composure. That was in 1979. Exactly a year later came a successful launch, and this time around, Dhawan declined Kalam's offer to address the press, ensuring that his younger colleague got his moment in the Sun after the spectacular success of his team's hard work.

There is also something to be said about the fact that space exploration relies completely on team work, and all teams assigned for a project must continuously talk to each other to ensure that they are in harmony and accomplish common goals through collaboration, for there is no way it can be successful without a team. This is another big factor that contributes to successful performance. In this conducive environment, the diamonds shine and sparkle. As a result, the awards and accolades have come home, such as the recommendations of the VI

Pay Commission of the Government of India that have given a significant reformative push to ISRO staff salaries, and the organisation stands out today as a case in point of sound people-centred policies that can otherwise be the greatest limiting step for a programme that relies completely on human endeavours of excellence. It must of course be said that the number of women in top leadership positions are not as many as there ought to be, which could partly be because engineering is traditionally a male domain. That said, there are women leading some of the teams at ISRO and about 20–25 per cent of ISRO's workforce is women.

K Radhakrishnan, Chairman, ISRO

In his keenness to master Carnatic devotional music, Radhakrishnan, ISRO's dynamic Chairman who has taken the Mars mission from dream to reality, has a sneaking feeling he has neglected his Kathakali in recent years. But, filling the void is a deep desire to excel

A man who started learning Kathakali, Kerala's traditional classical dance form, at an early age, K Radhakrishnan, Chairman ISRO, is seen here ready to perform (pictures from the mid 1980s).
Courtesy: K Radhakrishnan

The Mars man of India, ISRO Chairman K Radhakrishnan, a man of many talents who ably led his team to complete the mission within its exacting time requirements. Seen here against the backdrop of the MOM launch at Sriharikota.

Credit: Pallava Bagla

in something that he, in his own eyes, was mediocre at, and this is driving him closer and closer to music. Kathakali is a been-there, done-that sort of feeling, perhaps. In that intense, impassioned dance drama that Kerala is known for in the world of Indian classical dance, twilight has great meaning, as does the special kohl or *kajal* that burns the eyes till you blink away the tears and steel yourself for a performance, swathed in an elaborate costume and make-up. Radhakrishnan began dancing when he wasn't even 10 years old. This is familiar. I too started learning Kathakali at the age of six, always a reluctant horse kind of dancer, though. Nevertheless, I experienced intimately the intricacies of the form, each dramatic moment of those complex performances. Your body, mind, eyes, hands and legs are all stretched to their limits and you learn what it is like to be at the edge. That is perhaps why Radhakrishnan chases all his worries away when he dances. All the stress leaves him, he says, and it must be so.

The person who heads ISRO is not on a cake-walk. Everybody knows that. Right from

Seeking divine intervention. Dr K Radhakrishnan, Chairman of ISRO in the traditional 'mundu' that men in South India wear, and praying at the Tirumala Tirupati Devasthanams temple at Tirupati in Andhra Pradesh, not far from Sriharikota. Models of the Mars Orbiter Mission, the Polar Satellite Launch Vehicle and the official brochure of the Mars mission can be seen being given as offering to the Gods.

Credit: Jai Sekhar/Tirumala Tirupati Devasthanams

the Sarabhai and Dhawan days, the leaders of India's space programme have always taken pressure on the chin and smiled through it. They have stood up when there were failures, shielding their younger colleagues who have been on the shop floor, and stepped graciously aside when there are spectacular successes, allowing the teams and younger managers to take the bow. Radhakrishnan is the seventh Chairman of ISRO, and has a tough act to emulate if his predecessors are to be emulated. The longest serving Chairman was Professor Dhawan, who took over in 1972, after Sarabhai's sudden death in December 1971. For a few months, Professor MGK Menon held the post, after which Dhawan took charge. Radhakrishnan works hard at his job. Always has. And he remembers or knows no other way of being. Living with his banker wife – they have no children – Radhakrishnan came to ISRO as a young person, to take up his first job after passing out from the Government Engineering College, Thrissur in 1970, starting off as an avionics engineer at VSSC in Thiruvananthapuram in 1971. He was just 22, and spent the following four decades at ISRO.

Today, some quirky things give him great joy. When the Indian Institute of Management-Bangalore (IIM-B) announces its courses and guest lectures and he sees his picture there, among eminent speakers, Radhakrishnan is filled with a strange sense of fulfilment. Criticism and censure are natural companions on a leader's journey, and he has had his fair share. So, when

Carnatic music is a great passion for the ISRO Chairman, K Radhakrishnan, who still performs at temple festivals. Seen here at the Guruvayoor Devaswom Chembai Sangeetolsavam (2012).

Courtesy: K Radhakrishnan

ISRO's Chairman, K Radhakrishnan, accomplished musician and Kathakali exponent, giving a classical music performance in Antariksh Bhavan during the Karnataka Rajyotsava celebrations (2012).
Credit: ISRO

another institution of eminence, particularly from a totally disparate discipline acknowledges him, it makes him happy, he says. The management mantras that ISRO uses and operates with are certainly high-end; they usually deliver projects to sharp deadlines, and nurture and keep well and happy their 16,000 and more staff. It is therefore not surprising that they now have more successes than ever before, and of course a few failures. Radhakrishnan is a Distinguished Alumnus of IIM-B, and was part of the very first Masters in Business Administration (MBA) batch in 1976. He went on to also pick up a doctoral degree from IIT Kharagpur. The IIMs of India and ISRO are inextricably linked from birth – it was Vikram Sarabhai who created the first ever IIM of India, in his hometown of Ahmedabad, and he was the creator of ISRO too. Several mid-level engineers at ISRO are studying for their MBA at various IIMs in the country, and it is fascinating to see this close but complex link between the cultures and pulses of two great institutions of modern India.

Radhakrishnan believes he is still a student of music and has a long way to go before reaching perfection. He is a devout and practicing Hindu who visits the famous Sabarimala temple in Kerala that is an ode to Ayyappan, the male God worshipped mainly in southern India, who is believed to have been born of a union between Shiva and an enchantress Mohini. He also goes to Tirupati, a world-famous shrine in Andhra Pradesh, before every launch. Receiving

the Padma Bhushan in 2014 for Science and Engineering, especially in the field of Space Science and Technology, Radhakrishnan has come full circle, but doesn't believe in stopping in his tracks even for a moment. 'The day we got back after the Mars mission was successfully launched, I needed to go somewhere, but just as I was leaving I was told that the staff had got together to celebrate. They had decorated our foyer and I know that at such a time, even the canteen staff would be involved. So I stayed. That is how it is here, everybody takes pride in everybody else's work,' Radhakrishnan says with great pride in his institution, and since he has made a major leadership contribution, the pride is indeed appropriate.

M Annadurai, Programme Director, Mars Mission

When the Indian Moon mission, Chandrayaan-1, became global news, M Annadurai got a lot of attention, becoming in his own words 'a hero from zero'! His sudden and open laughter, his humility and the simple, straightforward way in which he explains complex space engineering projects endeared him to many. Here is a man who has had a long, long journey, from a small village called Kodavadi near Coimbatore in Tamil Nadu all the way to becoming the key technical expert on some of the Indian space programme's landmark projects in recent times,

The doer, M Annadurai, Programme Director for the Mars Orbiter Mission, for whom failure is not an option.

Credit: Pallava Bagla

first Chandrayaan-1 and then in just a few years, the Mars Orbiter Mission. These accolades sit lightly on a modest man's shoulders! Annadurai doesn't make much of it, passing off his responsibilities as just a day's work. He did not leave Coimbatore till his Masters' and is the man who now heads missions to the Moon, Mars, and who knows what other glamorous destinations in the future? It must not have been easy for the boy who grew up going to the Gram Panchayat school in his village, a village that did not even have electricity till he was a young boy. The lack of artificial lighting is perhaps what made the moonshine so special to Annadurai's childhood, and one of his most cherished memories is of his grandmother feeding him while showing him the Moon and telling him stories. To think that the natural planetarium the young, Tamilian boy gazed at each night and drew simple joy from would become his mind's playground in later years! There is something poignant about the fact that he helped discover water on the Moon even before the villagers in his home had the pleasure of continuous water supply. Even today, Annadurai and his wife live very simply in a small house in Bangalore, where the garage has been converted into a library that also houses his many trophies and prizes. It is evident that the drama is only in the actual work he does.

Annadurai is an accomplished space engineer known for his alacrity, his speedy decision-making and ability to deconstruct complex space engineering concepts and designs. He studied in Coimbatore till his post-graduation, and happily waves his arms to draw attention to his large room saying 'this room has come to me after almost 30 years of work' alluding to the fact that the expectations are not great, but the opportunities are tremendous. The only thing that matters to him is the quality of his work. He likes the culture at ISRO, with proper working channels from the top to bottom, so that people are able to connect with what they are doing, and feel the sentiments attached to their work. He believes all this has probably got carried forward from the time when Sarabhai and others built up what he calls 'graceful ownership' at ISRO. This tradition has 'instilled many values in us,' says Annadurai. 'Even during Chandrayaan-1, our international collaborators felt it was like working in a family. We are not bothered about sweating it out,' he says. 'For us Indians, one plus one is not two, it is three,' says Annadurai, who believes this is an important factor that has brought stories of success to the Indian space programme. 'It is like a hungry baby, whatever it gets, it will be happy, and this must have gone a long way.'

S ARUNAN, PROJECT DIRECTOR, MARS MISSION

PG Wodehouse books that were the zenith of wry, English humour through most of the twentieth century, and Bond movies, that's what it takes to make S Arunan a happy man. Actually, no. This is a man who is driven by the passion he has for his work, and from being the satellite manager that he is. But yes, when there's a moment of leisure, books and movies

S Arunan, Project Director of the Mars Orbiter Mission. He is believed to have slept in the ISRO Satellite Centre almost every day during the 15 months it took to build the satellite and get it ready.

Credit: Pallava Bagla

help him unwind. 'I completed my B.Tech in Mechanical Engineering from the University of Madras in Coimbatore and joined ISRO, it was my first job,' says the very genial lead of the Mars Orbiter spacecraft who believes much more in real work experience than in gathering degrees of higher education. When he was busy with the Chandrayaan-2 mission, Arunan got pulled into the Mars mission project, and feels very appreciative of his seniors for having thrown him a challenge, and a great professional opportunity. Arunan took it on in exactly the same spirit, and the 15 months till the spacecraft was ready to fly were crazy for this tall man who doesn't let the pressure show. 'I would go home only to perform my puja and freshen up, nights were spent right here at the satellite centre,' says Arunan. He was also very particular about being assigned young colleagues who would work with him, and that made the job not just more full with energy, but also fulfilling, with youngsters eager to learn and perform to their best. Meanwhile, at home, where his wife is a Montessori expert and his daughter currently studying to become a doctor, there is total support. His daughter, just a few years ago, brought him a Labrador puppy as a stress-buster, and the dog is very aptly christened Chocolate!

Arunan started working for ISRO at the VSSC in 1984, and the three decades that followed took him from one interesting piece of work to another. He sees great privilege in having been selected to head the Mars satellite management project. It is indeed uncommon for a mechanical engineer to be handed the task of designing and developing a spacecraft, the job that is normally the domain of electronics engineers. Now, after the MOM experience, he feels ready to take on anything, and take along a lot of the youngsters with him, on each exciting professional journey. 'There are so many advantages of working with young people. The team learns a lot, and so does ISRO, because young people 25–30 years old get trained and they are anyway good learners. They then are able to take forward the work for years to come.'

S SOMANATH, PROJECT DIRECTOR, GSLV MK III

Although not directly linked with the Mars mission per se, S Somanath, who is Project Director for the GSLV Mk III is a key figure for VSSC and ISRO. Starting at ISRO with the PSLV project in 1985, Somanath moved further, to integration, planning design, and implementing assembly at the launch pad, then going on to become a structures expert. At a time when it was most common for engineers from India to try to go to the US, and to top private industries, Somanath also thought similarly, but not for very long. In realising that VSSC was really the place to be, Somanath found his peace, and has never looked back since. The tall, strapping and striking engineer is also Deputy Director of the Structural Engineering Entity at VSSC. Spend a few minutes talking to him and you immediately start thinking that it is such a natural fit that he should be leading the glamorous GSLV Mk III – India's monster rocket project. 'My dreams were much greater than what I am today, but that is because dreams anyway have to be high. You should never doubt your capability. The excitement on the launch pad is a good feeling, that our baby is ready. That is how any person who does creative work feels, like a musician would feel after making a song. Is my work greater work than making a song? I don't think so, but it is neither routine nor monotonous.' Somanath believes strongly in the creative aspects of being a rocket engineer, and that the lack of bureaucratic structures gives them the freedom and motivation to be creative. 'People are very comfortable here,' he says, 'nobody acts like a boss, opinions can be freely expressed by all. At the same time there is basic discipline.'

Somanath sees himself as the consummate engineer, someone who likes to do something new each time. He truly believes that ISRO is able to achieve great heights in rocket engineering, among other things. 'Whatever we do here, we have tried and failed in private industry. So there is a sense of achievement, and we create through a process of analysis.' Needless to say, the recognition that comes with good work, and the promotions and encouragement to do something new and different, these are also critical factors that make people happy. 'I got my first appointment letter even before I completed my graduation,' says Somanath, who was one of the star performers in his batch at the Kollam College of Engineering in the early 1980s. Going on to get a Masters degree from the Indian Institute of Science in Bangalore, Somanath is also pursuing a doctoral programme at IIT-Chennai.

P KUNHIKRISHNAN, PROJECT DIRECTOR, PSLV

Even though the rocket he commandeers makes a deafening roar as it lifts off, P Kunhikrishnan speaks so softly you have to strain yourself to hear him clearly, and listen carefully as he speaks so you don't miss out on anything important! The gentleness is all-pervasive with this man who handles the PSLV, which in turn handles a huge bulk of ISRO's launches, placing craft after spacecraft successfully into orbit since it was first built in 1992. Completing his B.Tech from the College of Engineering, Thiruvananthapuram in 1986, Kunhikrishnan took up his first ever job with the Systems Reliability group at VSSC and then took on tasks for the avionics

On the rocket port, a cage for humans! Large parts of the island of Sriharikota are covered with arid forests that are home to leopards. When engineers travel around from one facility to another mostly on foot or bicycles, the instructions to them are to enter one such cage if they encounter a leopard by any chance. These cages are strewn around the entire area, placed strategically on the island.

Credit: Pallava Bagla

The quiet and unassuming P Kunhikrishnan, Mission Director for the PSLV. Behind the soft-spoken exterior is a man of exacting rocketry skills and precision that has taken the PSLV to great heights in the world of Indian rocketry.

Credit: Pallava Bagla

of all launch systems. After that he has been working on the avionics of all launch systems. Seated in his office on the ground floor of one of the VSSC buildings, Kunhikrishnan works with a group of about 35 core project members for the PSLV, on which he began with the PSLV C 12 to launch the RISat in January, 2009. 'Every mission, I learn so much, interacting with different people. For any project like this we have to approach people and appreciate their work. The seriousness with which they work, they have to feel they are contributing because even a small component can make the project go wrong. People should feel pride that we are doing the job for the nation. That is something special, not for our own selves or centres,' says Kuhnikrishnan.

This unassuming man never ceases to be surprised by the capacity that his organisation and its people have, not just for hard work, but also to remain self-motivated. 'Take my own experience, there was a time when the vehicle was fully integrated, but after some discussion, we had to de-stack the vehicle three times. We don't take any chances.' But nobody was complaining, and that is why he says it astonishes him, the way people work at ISRO. 'It is definitely the culture, right from the beginning, called the ISRO culture. In fact, the management side is equally motivating, for technical work, and for new ideas, and at the core of all this is team work that brings success.'

JN GOSWAMI, DIRECTOR, PHYSICS RESEARCH LABORATORY

Somewhat Einstein-like, wearing a slightly distracted expression on his face, JN Goswami walks casually towards me and settles down on the sofa in the unpreventious foyer outside his office. There is a languid smile that puts me at ease. He has devoted several decades in pursuit of excellence in space science and exploration, keeping an eye on India's scientific quest, while others at ISRO stay focused on the technology of travelling in the Universe. Goswami's passion is to grow the science of space, to get young people interested in the quest of planetary science through the Planex programme (Planetary Science and Exploration Programme) which was begun even before the Chandrayaan-1 mission, and to expand the base of people who can do such work. 'We have a school for post graduate and undergraduate studies, and we know there are sufficient people in the country who can interpret data for planetary science.' His stellar contribution with the Moon mission has been the role he played in interpreting data from the mission that led to the discovery of water on the Moon.

Goswami believes that the smartness of Indian engineering has got the country to this point where an entire inter-planetary mission could be mounted by the country single-handedly, without any collaboration. Chief scientist for the Chandrayaan-1 mission, Goswami often says about the MOM – 'This is a technology mission till it reaches Mars, and then it morphs into a scientific mission.' He of course feels that more space for accommodating more

The maverick scientist Dr JN Goswami, Director of the Physical Research Laboratory in Ahmedabad, who says Mangalyaan will remain a technology demonstrator till it reaches Mars, after which it will morph into a scientific mission.
Credit: Pallava Bagla

instruments to conduct greater experiments would have been outstanding for the Mars mission, but he isn't one to complain for too long. Passionate about asking tough questions that unravel scientific puzzles of the solar system and space, Goswami has seen Indian planetary science grow from baby steps to now. He began his higher studies at Cotton College, Guwahati, with

a B.Sc., and is today at the frontline of space scientists in the world respected as a top notch inter-planetary explorer. But all is not science, after all, and the man loves football, like most Indians, especially those who belong to eastern and northeastern India. But his wife has been known to say that his work is his life, and that he works 25 hours a day, 366 days a year. Seems entirely probable!

S RAMAKRISHNAN, FORMER DIRECTOR, VSSC

Leading VSSC, the cradle of India's space programme, is a challenging and tough task. Having been with ISRO for over 40 years, S. Ramakrishnan was completely focused on

The man who conducted early feasibility studies whether India could at all send astronauts into space. S Ramakrishnan, former Director of the Vikram Sarabhai Space Centre and accomplished rocketry engineer who has led VSSC from the front.
Credit: Pallava Bagla

getting the PSLV ready for carrying the Mars Orbiter, and his greatest challenge was to keep his teams motivated for the tight launch window, and the fact that they had to be ready in time for a November 2013 launch. Widely known as the architect of the PSLV, Ramakrishnan has been one of ISRO's leading lights. He was one of the early entrants at VSSC, joining APJ Abdul Kalam's team on the SLV-3 project in the early 1970s and never looking back for four long decades, while seeing the centre grow into a multi-disciplinary institution. Having studied at the College of Engineering, Guindy in Chennai, he also did an M.Tech in aerospace from IIT-Madras. Ramakrishnan received the Padma Shree in 2003 for his work and achievements. Ramakrishnan is the man who did some of the early work on the HSP.

AS KIRAN KUMAR, DIRECTOR, SPACE APPLICATIONS CENTRE

It would be best not to be misled by the small and slight frame of this ISRO stalwart. Honoured with a Padma Shree in 2014, AS Kiran Kumar is a distinguished scientist who finds that at the Space Applications Centre in Ahmedabad there is often a challenge that must be met. A space mission is designed and the SAC scientists have to retrofit their ideas of scientific equipment design and inquiry into what is available from the satellite designers in terms of size and weight. Kumar has perfected the technique of shepherding his staff to stay motivated, focused and in readiness to deliver as per the exacting specifications of their complex tasks. In getting the three payloads ready for the Mars Orbiter, miniaturisation was a major challenge, and the teams got focused on delivering the equipment to such exacting standards. Kumar believes that since all the work that ISRO does seems to have some end product or the other that is beneficial to society, and is therefore clearly visible in that sense, motivation to work comes along quite easily to most people. 'Something you have worked on is out there in deep space, what can be more thrilling than that?' he asks.

'We are hoping for the best and preparing for the worst' is the idiom that Dr Kiran Kumar, Director, Space Applications Centre, Ahmedabad goes by. He has provided leadership to SAC in the development of three payloads that are flying on the MOM spacecraft.

Credit: Pallava Bagla

Having studied at Bangalore University, Kumar also did an M.Tech from the Indian Institute of Science, Bangalore. Kumar joined ISRO in 1975, and has enjoyed every bit of the journey.

'The Mars mission was not a challenge, but a great opportunity,' is how Kumar sees the MOM experience. A great lover of all kinds of indoor sports, Kumar enjoys bridge sessions in Ahmedabad, and has also been an enthusiast of table tennis, badminton, chess and lawn tennis. Living with his wife, who he sees as the most understanding companion anybody could wish for, Kumar recalls how ISRO saw major staff attrition during the time of the internet bubble and follows the basic institutional principle that if people remain engaged with something or the other and are allowed to delve into new projects and take up fresh challenges on the job, they will remain interested in what they do. That is true, and stagnation is certainly one of the greatest bugbears of human resource management across the world.

M CHANDRADATHAN, DIRECTOR, VSSC

A man whose name itself has the 'moon' all over, M Chandradathan was heading India's spaceport when India headed to the Moon in 2008. Subsequently, he played a lead role at the Liquid Propulsion Systems Centre when the complex cryogenic engine technology was mastered. He now heads the country's main rokketry lab in Thiruvananthapuram.

The new Director of VSSC, M Chandradathan.
Credit: ISRO

MYS PRASAD, DIRECTOR, SATISH DHAWAN SPACE CENTRE

At the Satish Dhawan Space Centre in Sriharikota, action is a daily pattern. But, the unflappable MYS Prasad, an old ISRO hand who is also Chairman of the Launch Authorisation Board was cool and calm when he was told that the Mars mission launch would be in October-November, a time when there is cyclonic weather on the eastern Indian coast, during the northeast monsoon season. The record is daunting – Prasad has seen ISRO through 19 launches at Sriharikota, 16 of the PSLV and three of the GSLV. Having joined

MYS Prasad, Director of the Satish Dhawan Space Centre at Sriharikota. He was instrumental in averting what could have been a major calamity at Sriharikota when a serious fuel leak involving about 750 kgs of highly inflammable rocket fuel got detected just hours before the lift-off of India's Geosynchronous Satellite Launch Vehicle (GSLV) Mk II.
Credit: Pallava Bagla

ISRO on Kalam's SLV-3 team, Prasad says it is a 'neck into the noose' kind of life that he leads, speaking of the rockets he gets launched. Shouldering the massive responsibility of running Sriharikota, India's rocket launch port, it is Prasad who detected a fuel leak on the GSLV once in 2013, and his sharp eyes picked what could have been a near disaster for India's space programme as the vehicle and launch pad could all have blown up. Every second counts for Prasad on a daily basis, and just managing large rockets with such large quantities of propellant, all filled up and ready, is no cakewalk. No wonder he says, 'Tomorrow I may not be there, but today, we have to work with operational discipline and technical democracy.'

A major challenge for Prasad is the fact that at the top, one has to have complete knowledge of all aspects of each operation. 'In my second role as the Chairman of the Launch Authorisation Board, all responsibility lies with me,' says Prasad. But, as he says, ISRO leadership worries more for the organisation and hardly works to please anybody, so this gives a lot of clarity of purpose. For a man who works at the deep end, quite literally, handling ISRO's biggest machines and most critical operations, Prasad indeed must value that.

SK Shivakumar, Director, ISRO Satellite Centre

ISRO's heart beats in his centre, where satellites are created and crafted as though they are just some ordinary pieces of machinery. But, one thought of the kind of scientific and technological applications that satellites are used for, and it is easy to balk at the kind of responsibilities SK Shivakumar shoulders on a day-to-day basis. Known as the antenna maker, responsible for the 32 metre dish, Shivakumar hails from Srirangapatna near Mysore, the land of Tipu Sultan, the original rocketry man of India. In fact, the first initial of his name stands for Srirangapatna. With a confident, clipped style of speaking, Shivakumar clearly states that every satellite that is made is special, and each and every system is followed rigorously in the making of satellites. The Mars satellite was no different, except for the tight timeline it had to be completed in. 'I'm just one job old,' he says jokingly, of the fact that he joined ISRO for his first job, and never left. Of course, the support from his wife and two daughters, both of whom have followed

SK Shivakumar is the Director of the ISRO Satellite Centre in Bangalore. Shivakumar hails from Mysore, the ancient cradle of Indian rocket science, led by the inimitable Tipu Sultan. Shivakumar led the team that fabricated the 32-metre dish antenna at Byalalu.

Credit: Pallava Bagla

in their father's footsteps and become engineers, has been an immense contribution to the fact that Shivakumar today feels good about how his career has progressed.

'We have been taught at ISRO to work in project mode. This helps keep up the motivation tremendously, because then there is diversity in work that each one does,' says Shivakumar. What keeps him going is still the same old passion: 'When you build a system in the lab, and then you see it working in space out there, that thrill just cannot be described, and that is what keeps us going.'

DR SUMA, GENERAL MANAGER, I-SITE

One of the few women at the helm of affairs, DR Suma goes about looking after her domain with a cheerful smile and a certain air of pleasantness. A graduate of the Government Engineering College, Thrissur, the same college from which India's missile woman Tessy Thomas graduated, Suma joined VSSC in 1983 as a quality control engineer, and as is most common with this organisation, kept at it till the current when she is looking after one of the most critical hubs of the space programme, the facilities that integrate different components to make them mission-ready. 'I doubt if there are many people at ISRO who have had such a

Woman power at ISRO. The pleasant and knowledgeable DR Suma, who is in charge of the ISRO Satellite Integration and Testing Establishment Centre (I-SITE) at Bangalore.
Credit: Pallava Bagla

wide gamut of experiences. I started working on launch vehicles, then moved to management at headquarters and am finally managing satellites,' says Suma, who believes that the sky is the limit for those who work at ISRO. Suma is fond of saying that a lot of what she learnt at work is owed to G Madhavan Nair, former Chairman of ISRO and earlier Director of VSSC when Suma worked there.

The former dancer – Suma has learnt Bharatanatyam and Mohiniattam – believes that any woman who is a professional has to prove her skills and her talent much more forcefully than a man has to. She attributes her success to her husband and son who have been her constant support. Suma is an avid practitioner of yoga, particularly since she discovered a chronic back problem that needs to be kept in check. Space administration is a passion, as is managing I-SITE, where, at any one point of time, there can be at least seven or eight satellites on the floor. The woman who emphatically refutes the idea that Mars is for men as ridiculous, saying that it is equally for women, goes around her work site with a high voltage smile and easy attitude that belies the stressful responsibilities that come with her position. The music and dance, of course, are stress-busters.

Dr S Unnikrishnan Nair, Project Director, Human Spaceflight Programme at VSSC

'It has been 28 years, and not once has it crossed my mind that I could be doing something else, or quit my job and take up another,' says the still very young-looking Unnikrishnan Nair who is now holding a significant portfolio for ISRO, that of the human spaceflight initiative. Completing his BTech in 1985, Nair was just 22 years old when he took up his first job with ISRO. Even today, his face still shines with youthful enthusiasm and excitement as he speaks of the work they do. He started off with the Aerospace Mechanism Group at Valiamala in Thiruvananthapuram, and the first years were spent on a journey that initiated the PSLV in India. 'The amazing thing about working at ISRO is that you get to hone your professional skills and talents as you go along,' says Unnikrishnan, who like many other colleagues, has studied alongside an active career. He did an MTech at Indian Institute of Science in Bangalore in 1992 and then went on

S Unnikrishnan Nair, Project Director of the Human Spaceflight Programme at VSSC.
Credit: ISRO

to do a PhD at IIT-Chennai. He is very fond of plants and Kerala, with its natural, tropical greenery and high rainfall, is of course a haven for plants.

BS Chandrashekar, Director, ISRO Telemetry, Tracking and Command Network (ISTRAC)

Listening to the whispers of the Mars mission, Chandrashekar has a tough job to keep. Handling the entire tracking network for an inter-planetary mission where one communication signal could take about 20 minutes one way, the whispers hold a special place in Chandrashekar's life. India's largest dish antenna, the giant saucer that is 32 metres in diameter, and placed at Byalalu off Bangalore, is his ear to the Mars mission.

BS Chandrasekhar, Director of the ISRO Satellite Telemetry, Tracking and Command Network, ISTRAC, in Bangalore. The man who does not want to ever hear MOM saying, 'Bangalore, we have a problem'.

Credit: Pallava Bagla

V Adimurthy, Satish Dhawan Professor and Dean of Research, ISRO

This is the man who is possibly the true-blue Martian explorer at ISRO! He was tasked with preparing the very first internal report to determine the feasibility of ISRO's abilities to undertake an inter-planetary mission to Mars. One of Adimurthy's major responsibilities today at ISRO is to monitor space debris, coordinating with global space debris monitoring organisations. He decides if a particular launch needs to be delayed or held off by a few minutes if there is any space debris coming in the way. In fact, the PSLV launch that was witnessed by PM Modi on June 30, 2014 was delayed by three minutes for exactly the same reason. In his earlier avatars, Adimurthy's was the voice that the nation would hear when the ISRO launches took

V Adimurthy, a man tasked to track space debris for ISRO. He wrote the first internal report for the Mars Mission. On most missions in the past, his was the voice that the world heard through his live commentaries that would accompany ISRO launches.

Credit: Pallava Bagla

place, as he was ISRO's special commentator in English giving a live account of the progress of any rocket launch from Sriharikota.

It must be said that one doesn't meet as many women as one would like to, particularly when ISRO seems to be steeped in southern Indian cultures. Less than a quarter of ISRO's workforce is women, although it is known that Indian engineering colleges have a growing number of girls. This is even more evident among senior leadership. Of course, ISRO is quick to dismiss the observation, but the gap is quite obvious. This could well be because few women choose to become rocket engineers or space scientists, but the clean rooms at SAC are filled with young, women scientists. Organisational trends have often revealed that despite the fact that women are in good numbers in the lower hierarchies, there is a petering out that happens as they move up the ladder. At ISRO Headquarters, meeting women leaders doesn't give one that sense, but for an organisation as mature, there ought to be greater visibility of female leadership. For example, ever since its inception in 1969, ISRO has never had a single woman as Chairperson, despite being among the largest government space agencies in the world. But the women one meets are not complaining! 'ISRO was my first job, and my dream job,' says TK Anuradha, an electronics engineer who is India's first woman satellite fabricator heading the GSAT-12 as Project Director. The continuous demand for transponders throws plenty of challenges Anuradha's way, but she thoroughly enjoys her work. Originally from Shimoga, Anuradha studied in Bangalore from middle school on, and considers herself a Bangalorean. 'We are into very brisk work, doing some seven to eight satellites every year, while at any point of time there are at least 15 projects going on at the centre. In 2013 we did six satellites, and more are lined up. I work with 20–25 Deputy Project Directors, with one or two project managers under each,' she says. Working for grass-roots community requirements is Anuradha's passion, and her belief is that delivering health and protecting lives from disasters, managing forest fires and information for villages, besides e-governance and automatic weather monitoring is something that satellite communication has managed to do increasingly well in the last few decades. Indeed, that is true. 'But we need to worry about sustaining these practices, because ISRO cannot take it to the last mile,' is what Anuradha feels. Having begun her professional life with assembly integration and testing, Anuradha first came to ISRO by responding to an advertisement in the newspaper. After all these years, she is almost one with the ISRO culture. 'We don't feel any difference as women, it just doesn't matter. I think there is a culture in ISRO, it is something fantastic; there is a lot of respect for one another. It's a beautiful place, with no serious allegations of marginalisation or anything like that.' She of course gives a lot of credit to her family too, accommodating as they are to her work and travel schedules. 'My parents-in-law supported me a lot throughout my career,' says the satellite woman. She is equally conscious of the fact that it depends as much on how a woman grooms her family

The 'eye' of the Mars mission, Ashutosh Arya is the engineer behind the Mars Colour Camera that is on Mangalyaan. He is part of the Space Applications Centre team at Ahmedabad that is responsible for the payloads.

Credit: Pallava Bagla

Sniffing for methane, perhaps the most exciting question that the Mars Orbiter Mission is asking. Kurien Mathew is the engineer from the Space Applications Centre, Ahmedabad who has led the project designing the Methane Sensor for Mars.

Credit: Pallava Bagla

Does Mars have fever? Checking for thermal signals will be the Thermal Infrared Imaging Spectrometer (one of the payloads on the MOM spacecraft) designed by RP Singh of the Space Applications Centre in Ahmedabad.

Credit: Pallava Bagla

Dr Anil Bhardwaj is the designer of the Mars Exospheric Neutral Composition Analyser (MENCA) payload, which will actually taste Martian atmosphere and try to unpack its composition. Bharadwaj was also involved with Chandrayaan-1.

Credit: Pallava Bagla

Deciphering the mystery of the absence of water on Mars and its presence on Earth, M Viswanatham is the man who has designed the Lyman Alpha Photometer for ISRO. This instrument will study the presence of different avatars of Hydrogen on Mars.

Credit: Pallava Bagla

and builds a support system that takes care of the home when she is away. 'Maybe I am just lucky, in my family there is total respect for each other's work.'

It is indeed fascinating that most women who reach leadership positions make a lot of effort to openly acknowledge family support, and one dare say such acknowledgements do not come as spontaneously from men in similar or higher positions. But these are sociological patterns that we live with and accept and conform to. For N Valarmathi, Heading the Digital Systems group as Group Director at the ISRO Headquarters, a similar thought process is triggered when one asks her about what it feels like being a woman leader within the Indian space programme. 'First, I thank my family for putting me into a professional college,' she says. Every project is special for this woman who has been around for three decades, viewing the vast programme with its many challenges, 'where every piece of work has to be useful to the common man, has to be goal and schedule oriented'. Valarmathi believes that these are the factors that drive ISRO to do things efficiently and effectively. Having worked on communications satellites in the early days of her career, Valarmathi recalls how it was challenging, because many of the women had small children then, sometimes even a few months old. Projects, once they get into full swing, can be very demanding in terms of timelines, rigour and accuracy among other things. 'Even if you work 24 hour a day, safety is not an issue at all,' says Valarmathi, continuing, 'Whether an engineer, an executive, or a technician, ISRO gives very big responsibilities to women. A piece of work may end by 3–5 am, with a technician who has to complete one last procedure of closing a panel. There is a procedure that has to be done that way. If you encourage women, they are mature, they are calm and they are patient, so that benefits the organisation and the country. This is my faith. Now so many young ladies are coming forward to work, so we should be setting an example.' Meanwhile, Geetha Varadan defies several commonly-held notions since she heads the group that interfaces between ISRO and the defence forces through a unit called ADRIN, the Advanced Data Processing Research Institute, in Hyderabad.

When the Indian space programme was first conceptualised, Sarabhai often saw people at the centre of his planning process. A lot of the great work environment for ISRO staffers is because that early thinking has been kept alive, like a warm fireplace. After all, people prefer to work with leaders than with managers, for the latter are perceived as controllers whom one should fear, while the former work through leading by example, not by coercion. When the young team that handles the ISRO Mars mission Facebook page, young staffers just at the beginning of their careers, walk about with a calm and happy air, something must be right. To them, ISRO is an informal place and that makes them happy. They are encouraged to question, to get into problem-solving and to also grow within themselves the autonomy of decision-making. 'If my strengths are identified early on, I have a chance to grow into my dreams,' says Shamsuddin. It is a similar feeling for young scientists at SAC, where each time a challenge is presented, posing the need for some technical development, there is great satisfaction in meeting that challenge, for it is much more than just a production kind of thing. Also, the variety of work that is presented as an opportunity is a driver of motivation. At

The Moon man of ISRO, G Madhavan Nair, Former Chairman, under whose leadership India made its maiden voyage to the Moon. Seen here holding an outstanding three dimensional image of the Moon. He subsequently fell afoul of the current ISRO establishment, singed as he was by the alleged S-Band transponder scam involving the Antrix-Devas deal.

Credit: Pallava Bagla

Celebrating India's arrival on the Moon, on November 14, 2008, the national flag was sent hurtling down from the Chandrayaan-1 satellite onto the surface of the Moon, using the Moon Impact Probe (MIP). Seen here cradling a globe of the Moon depicting the flag are (L to R): G Madhavan Nair, then Chairman of ISRO, former President of India APJ Abdul Kalam and TK Alex, then Director, ISRO Satellite Centre, K Radhakrishnan, then Director, Vikram Sarabhai Space Centre. UR Rao, father of the Indian satellite programme and former Chairman of the Space Commission, seated (extreme left), looks on.

Credit: Pallava Bagla

UR Rao, former Chairman of ISRO, who led the group that selected the scientific payloads that would fly on the Mars satellite. In hindsight, he says India should have attempted a mission to Mercury instead, which would have really pushed the envelope of space exploration.

Credit: Pallava Bagla

SAC in Ahmedabad, where a group of scientists who are Principal Investigators for the three payloads on the MOM gather, we get deep into discussion about these very drivers. For Dr Ashutosh Arya, the man behind the Mars Colour Camera, who studied geology at the Masters level from the University of Baroda (now Vadodara), the last 27 years have been with SAC, and he has enjoyed each minute. 'The heterogeneity of the nature of applications of our work and the projects that we get to face and work on is really tremendous. What if I were in the oil industry? I would always have been doing just one thing all my life. Here I have studied the Earth, peeped into resources like ground water and surface minerals but also worked on planetary systems, which people do not get to study in their entire lifetime.' What more can anybody ask for? Dr RP Singh, who is PI for the TIS, and Kurien Mathew for the MSM, echo the same sentiment, as does Somya Sarkar who is the Associate Project Director for the payloads and their development. M Viswanatham has likewise developed the LAP, and works in Bangalore. Dr Anil Bhardwaj, who is Director of the Space Physics Laboratory at VSSC and is the person behind the MENCA payload, believes that the ISRO job gives him everything – creative satisfaction, the freedom to ask challenging scientific questions and the pleasure of seeing end results.

For scientists who see the power of their work as it rolls out in outer space, life is somewhat different from the world of pure science. As John P Zachariah, Distinguished Scientist and Director R&D with the VSSC, puts it 'I was just a B.Tech, but am now more learned than a Ph.D. I go to international conferences and represent my organisation on committees, so it is more about how we work.' In running this massive organisation, some principles are applied as non-negotiables, like the reviews and discussions where there is free flow of expressions, a freedom much valued by all, because nobody is made to feel small, regardless of what perspective or opinion is shared. Technical openness is held dear at ISRO since every design and concept goes through rigorous technical reviews. In all these committees, anybody who has background knowledge and has a question or a suggestion can vocalise it, because the culture of questioning, providing positive criticism and feedback for improvement of design functionality, and acceptability is respected and valued. Failures are analysed threadbare and failure analysis committees literally take projects through a wringer. There is tremendous clarity on what needs to be done for each person. Fault finding is not there, and the team approach overwhelms everything. Stagnation, that greatest enemy of any organisation, is a word unheard of at ISRO, with the kind of movement and cross-team alignments that every person is directed towards. Meanwhile, the work schedules and project goals remain laser sharp, undiluted by individual ambition, and this lends a somewhat missionary approach to the space programme. As Radhakrishnan says, 'What motivates people is that they see concrete results of their work, and there is visibility of the same.' After all, it is the end result of work that gives purpose to people's lives. Radhakrishnan also feels that young people see role models among their seniors and note how 'there is no place for complacency. Of course self-actualisation has to happen,' he adds. There are few here who work for the money. There is also recognition, with a promotion structure that is quite different from other government departments, and specific criteria of review. It is well-known that former Chairman Nair worked hard to get the recognition and reward system in place and this has made a huge

One of the key ISRO leaders who was instrumental in making the Chandrayaan-1 satellite, Dr TK Alex was former Director of the ISRO Satellite Centre in Bangalore and conducted periodic reviews for the Mars Orbiter Mission.

Credit: Pallava Bagla

President APJ Abdul Kalam, the aeronautics engineer who mastered rocketry for India.

Credit: Pallava Bagla

difference. Somanath says, 'Our healthcare system is very good. Transport is very good. We are taken care of, in terms of something to eat, dropped home. You are prompted to work more and deliver more, that environment is created.' As P Ratnakara Rao, Deputy Director at VSSC says, 'There is a pull from the top to perform, and an equal push from the bottom to do the same.' This is shaped by leaders like V Koteswara Rao, Scientific Secretary for ISRO, who frontends a lot of the scientific decision-making at ISRO and was the first Project Director for Astrosat, India's forthcoming observatory in space.

Vikram Sarabhai is known to have been a great supporter of individual excellence in contrast to many leaders of institutions who become enslaved to hierarchies and organisational charts. He believed that no organisation should give too much importance to mere years of experience, and no organisation chart should stand in the way of recognising and rewarding talent. In a sharp observation Aristotle had once said, 'Excellence is an art won by training and habituation. We do not act rightly because we have virtue or excellence, but we rather have those because we have acted rightly. We are what we repeatedly do. Excellence, then, is not an act but a habit.' ISRO seems to have imbibed this philosophy, and working with young people from seemingly mediocre backgrounds and small-town colleges and institutes, has turned itself into a machine that performs like there is no tomorrow. Today, with the Universe to discover, this people-machine is on a roll, and with the younger lot getting into managerial positions, there is a lot to be excited about. Astronaut selections will come up with the human spaceflight programme taking shape, and these would obviously be special people with military aviation backgrounds, and the programme per se would need a mix, some with such backgrounds and others with the typical, ISRO scientist profile. Whatever the case, expansions are on. Radhakrishnan remains cool and has the confidence that the ISRO spirit catches everybody, always. He says, 'Have passion for whatever you do. Remember, perpetual optimism is a force multiplier.'

Chapter 7

International Explorations and Global Plans in a Shrinking World

The deep Indian summer of 2014 was a summer to remember. A new Prime Minister took office after a landslide win and dramatically called out to India's space community with his unusual use of a popular tag line made immortal by the valiant young caption Vikram Batra during the Kargil war: '*Yeh dil maange more*', meaning this heart wants more. Strangely, this is what seems to be ISRO's mantra also, moving tirelessly as it seems to from one project to another, covering decades and distances in space. Playing host to five foreign satellites going up on the trusted Indian Polar Satellite Launch Vehicle, Narendra Modi clapped in appreciation and then made a scintillating speech, connecting with his audience in a full room at India's space launch pad in a way one has rarely seen any political leader do in the recent past. Impressive orator that he is, the Prime Minister suddenly mentions a SAARC satellite, saying that should be ISRO's next challenge: 'Today, I ask our space community, to take up the challenge, of developing a SAARC Satellite – that we can dedicate to our neighbourhood, as a gift from India. A satellite, that provides a full range of applications and services, to all our neighbours. I also ask you, to enlarge the footprint of our satellite-based navigation system, to cover all of South Asia,' In the charged environment that envelopes Sriharikota each time there is a successful launch, are these the shifting sands of global power equations? Not so long ago, in the grips of the Cold War, the US and Soviet Union kept going to space – firing salvo after salvo. But in today's massively equalising world, space is more open and roomy, and will slowly accommodate more and more nations (even dark horses like Israel) and collaborative ventures, including private ones. So, Prime Minister Modi's SAARC satellite may head off sooner than we might think.

So may other exciting partnerships. In a totally different part of the world, in Israel, space exploration is serious business. The city of Tel Aviv, as I go about my day, tells me that in more ways than one. Stray thoughts shoot through my mind. Given my wanderlust how had this tiny nation never featured on my list of must-visit places? Just like India, here too the year 1947 was special and unforgettable in a million ways. India found freedom after almost 200 years of foreign rule by the British in the month of August, and Israel was actually founded as a new country in November that same year through a United Nations Partition Plan for Palestine. India has learnt to live and grow into a country to be reckoned with on the world stage despite some really hostile neighbours, so has Israel. And of course, just like it is at ISRO, here too one goes through several layers of security at the Space Head Quarters in Tel Aviv, to view Israeli satellites in the making. This contrasts somewhat with the Chinese, who also have big space ambitions. In that vast country that has spread its manufacturing tentacles across the world, I find some of their scientific institutions working on space exploration located right in the middle of the global city of Beijing, easily accessible provided you have an invitation from the Chinese government.

Meanwhile, the big muscles of the US and its many years of highly advanced and expensive space exploration pulse out of the rather nondescript matchbox-type buildings of NASA's headquarters in Washington DC in the capital district. As we walk in, I cannot but notice that even the NASA administrator has to flash his identity card at the entry. No special treatment by the security guards. For me, used to the Indian way of things where senior officials and many others are given special treatment almost everywhere, nicely iced with a sharp *salaam* from a suddenly alert guard, this image of General Charles F Bolden, also a former astronaut, going through the paces like any other person, endures forever. These little things distinguish the US and make it what it is today.

But all differences aside, space is the great equaliser of nations. It is truly the vast and dark and unexplored Universe that envelopes all of us, regardless of privilege, riches or creed. And that is perhaps what Vikram Sarabhai was thinking when he arranged for a global galaxy of stalwarts and experts to view the first rocket launch from the beaches of Thiruvananthapuram, gesturing to American collaborators that India would fly the Nike Apache rocket they had lent. Almost repeating the thought behind that history, Bolden said in a wonderfully wide-ranging and enjoyable interview, 'Two democracies, the largest and the oldest in the world, and I think it is important that we show that democracies get things done much more effectively than other forms of government, and the fact that with common goals and aspirations, the world's oldest and largest democracies can work together.'¹ Who can deny the nobility in the idea? Its practicality, however, is often fraught with unimaginable challenges in the real world of today. More often than not, competition has edged out collaboration, and space races have outshone international team-work.

Mars is on everybody's agenda. It's a crowd out there, led now by the new kid on the block, the US's Curiosity, the new generation rover that landed in the Gale crater in 2012 and has since been trekking all over the planet, completing one Martian year (687 Earth days) in June 2014. Curiosity joined the US's Opportunity and Sprit on the planet, and orbiting around Mars are quite a few companions. The 2001 Mars Odyssey is still capturing pictures, the Mars Express and Beagle 2 have been around for more than 10 years, the Mars Exploration Rover Opportunity crawled about till it got to the 22-kilometre crater called Endeavour, and the Mars Reconnaisance Orbiter is checking for water and using its powerful camera to determine future landing sites for other machines. The new entrants, India's MOM and US's MAVEN, join this family in September 2014. So that's a mélange of the US, ESA

An artist's impression of the Mars Opportunity Rover which has been active on Mars since 2004.
Credit: NASA

and India, out there together—and it has a nice sound to it. It seems like competition in small doses is not a bad thing after all. This aphorism perhaps rings the truest in the world of space science research, exploration and technology. Studying the Universe is big science and the very idea engenders big thinking, collaborations and partnerships, because it all seems to make sense. Somehow though, laden as it is with strategic and defence-related undertones, space technology and its use for exploration has seen international collaboration from nations that have the capability only in a highly measured, often just symbolic and cautious sort of way. In the contemporary world, while collaboration is clearly the best way forward, institutions, led by governments and nations, have remained quite narrow-minded and out of sync with modern times, each continuing to pour in precious resources for space exploration, without much thinking about reaching out and partnering on such ventures. Or even sharing the spoils that include data from research and exploration, except commercially of course. At the same time, while some data may be classified and not shared, what we all know today about Mars is not because every country mounted its own inter-planetary missions, which is anyway not possible, but because the open source of knowledge kept getting built over time. Despite the fact that nations are generally very closed and possessive about most of their own programmes and initiatives, resulting in a lot of replication of work, what cannot be denied to any person is the wealth of information and data that is contributed to a global pool of knowledge whenever any nation plans and conducts a spatial mission. That begs the million dollar question – why not partner and collaborate when in the long run everybody is anyway helping grow a worldwide knowledge pool? This is especially important to understand against the backdrop of international frameworks like the UN

Mars Odyssey, an artist's impression. Odyssey is a satellite orbiting Mars launched by NASA back in 2001. One of its main purposes was to answer if life existed on Mars or not.

Credit: NASA

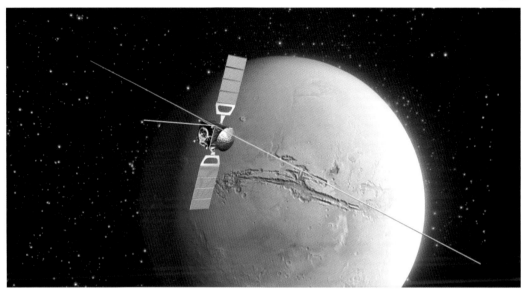

Mars Express, the European Space Agency's satellite orbiting Mars. Launched in 2003, it continues to provide spectacular views of the red planet. An artist's impression.

Credit: ESA

General Assembly's Resolution 1721 (XVI) related to International Cooperation in the Peaceful Uses of Outer Space that dates back to 1961. The Resolution clearly lays out the need for international cooperation for the benefit of nations 'irrespective of the stage of their economic or scientific development.' International covenants aside, India, as is to be expected, has its own policy frameworks and commitments that underscore the need for international collaboration. During the eleventh Five Year Plan period (2007–12), these commitments saw tangible results in the Chandrayaan-1 mission where six of the 11 payloads were from countries other than India. In fact, the Report of the Working Group WG 14 for the twelfth Five Year Plan highlights this achievement of a multi-lateral cooperative mechanism allowing 'significant progress in international cooperative endeavours and initiatives.' In a similar vein is cited the Megha-Tropiques satellite as a 'shining example of Indo-French cooperation.'

Prime Minister Modi, in throwing the SAARC satellite challenge into the ring, seems to have aligned himself quite closely with the existing policy articulation in the Space Commission's documents, which refer not only to the need to continue to strengthen and establish new relationships of cooperation with space entities in other countries, but also specifically calls for ISRO to help establish a network of weather stations in SAARC countries to support severe thunderstorm predictions and manage disasters. India's plan commitments include the need to enhance the presence of Indian space capabilities on international forums, ideas to jointly develop and launch small satellites with the United Kingdom, share satellite data with the Association of South East Asian Nations (ASEAN) for disaster management support, take up joint activities in space science under the aegis of the International Space Exploration Coordination Group (ISECG), and work with ESA, NASA, the Japan Aerospace Exploration Agency (JAXA) and the Centre National d'Etudes Spatiales (CNES) to realise satellites for Earth observation and Earth system science. India is expected to continue to share its expertise in capacity building and disaster management support with needy countries through various international fora. ISRO expects to continue to actively participate in activities of the Global Earth Observation System of Systems (GEOSS) and also support space-based virtual constellations of satellites for various themes by committing its satellites and data products for the global cause.[2]

The world is changing, and fast. India is now being viewed as a 'safe partner by developed nations and a leader for developing nations'.[3] In fact, as part of this evolution, India may choose to collaborate more as a goodwill gesture than out of necessity. Mangalyaan, the low-cost speed runner from India, has attracted global attention from the international space community for a variety of reasons. MOM is a completely indigenous mission except for some stuff picked up from the international space technology market place. ISRO claims there was really no time to reach out to any international partner. Meanwhile, it is obvious that India has developed end-to-end capability. There is also distinction in the Indian space programme's delivery against shoe-string budgets, in global comparison. The annual ISRO budget averages about USD 1 billion, contrasting sharply with US budgets, where NASA's Mars Curiosity rover alone cost about USD 2 billion. It isn't as though the work is less than gold standard. Even Bolden was impressed with the Indian space programme's facilities: 'Recently, I went to Delhi and then to Ahmedabad. It was incredible to go there and look at their [Space Applications] Centre and satellites that were under construction and the missions that were in planning and to look at the commonality between the things that we do in the field of science, in environmental science, looking at water issues. I was very impressed with the facilities, I got a chance to see two or three different clean rooms and three different types of missions under construction, it was an impressive operation.' India did lead from the front when the Moon mission was planned and launched, and since it was an unusual time with the country under sanctions, this leadership automatically brought with it the distinction of a developing Asian nation attracting partners from seasoned, advanced space-farers like the US and European nations. Chandrayaan-1 has become something of an exemplar in how a number of nations were invited by India to fly their scientific payloads on the Indian satellite and as Bolden has said, 'If you look at what we have done co-cooperatively, India and the United States participated together on Chandrayaan-1, orbiting the Moon, and getting data from there, Chandrayaan-1 discovered a significant amount of water on the lunar surface'. This would certainly be the ideal way to explore the Universe, but it must be said that collaboration is certainly not what distinguishes the more than 5,000 missions involving either near-Earth satellites or deep space travel that have taken place in the last six decades, broadening beyond imagination the human understanding of the Universe of which we are a part.[4] What was a dark and unfathomable amorphousness that teased human curiosity by its enveloping nature is today an unravelling composite of skies, planets, stars and many other elements of nature. The mysteries of the cosmic Universe may never cease to tantalise us, but the familiarity has grown in equal measure. This large body of knowledge has been built up over time, and is the contribution of several nations and their programmes.

In the early years, the US and the erstwhile USSR were running a crazy space race. Global political messaging played a significant role and there were many times when the world knew that a particular space mission had been

The woman who found water on the Moon. Carle Pieters is a professor at Brown University in the US who flew the Moon Minerology Mapper on Chandrayaan-1, and this is the instrument that gave the first evidence of the presence of water molecules on the Moon surface. It was this finding that propelled India onto the global interplanetary map, catalysing the country's decision to go all the way to Mars.

Credit: Pallava Bagla

launched less in the quest for knowledge and more to transmit a clarion call of supremacy. Geopolitics, the strategic positioning of space-faring activities and the complex shades of grey that get superimposed because of the close alignment between peaceful and military intent when it comes to such activity, all of these factors have shaped what the international setting is like today, as far as Mars and other space missions are concerned. But more about that later. Meanwhile, it is the politics of space that might be behind how several early Mars missions were failures. If a world museum of inter-planetary missions was to be created, it would automatically train the spotlight on how a small clutch of countries has continued to try and try again, despite failures, because there is nothing as heady as the sweet taste of success (see

Annexure 7). Equally liberating is the feeling that comes from proving one's supremacy on the global stage; and in modern times when few real wars are fought to establish the ruler, space wars are quite the rage.

As is quite understandable, early efforts of the 51 missions mounted between 1960 and 2014 were generally fly-bys, then graduating to orbiters and landers. The Korabl 4 USSR Flyby was the first of six fly-by missions that began in 1960, and the Russian superpower spent the next ten years from 1960 sending mission after mission, with not a single success. The US also sent one at the time which was a failure. They just could not reach the required altitude. Then, half a century ago, Mariner 4, a small robotic explorer weighing less than a ton, was sent by the Jet Propulsion Laboratory (JPL) in the US and got written into history as the first successful mission to Mars by any country. In 1971, the USSR managed to notch its first success, with the Mars 3 Orbiter/Lander. All in all, there were 18 missions from the USSR between 1960 and 2012 and 21 from the US, between 1964 and 2012, with the USSR having two successful missions, and a third partly successful, and the US having five successful missions. The Japanese tried getting to Mars but once, with the Nozomi in 1998, which failed. ESA sent the Mars Express Orbiter/Beagle 2 Lander in 2003, which was a partial success. The European Orbiter managed some great and detailed imaging of Mars, but the Lander was lost on arrival. Then came the late nineties and between 1996 and 2007, seven major NASA and ESA planetary spacecraft successfully achieved either insertion into the Mars orbit or landing on the red planet's surface. These missions, coupled with continuing analysis of Martian meteorites, have generated an enormous wealth of new information on the search for evidence of life on Mars.[5] But it was really the Mariner 4, setting out on November 28 in 1964 that presented to the world the tantalising possibility that humankind was capable of exploring other planets of the Solar System. It took the first close-up images of mysterious Mars in July 1965, after an eight month long voyage, then stayed in orbit for about three years, continuing to study the solar wind environment and making coordinated measurements with Mariner 5, which was sent to Venus in 1967. The US Mariner Series between the mid-1960s and the mid-1970s significantly upgraded our understanding of Mars. Mariner 4 came back with information that Martian air is almost nothing but carbon dioxide (CO_2) (95 per cent in fact). It also told us that atmospheric pressure is so low on Mars that liquid water cannot exist on the surface. The Hellas basin, which is one of the best known regional features of the Martian surface, was discovered by Mariner 6, while Mariner 9 managed much more – Olympus Mons, Valles Marineris, canyons, the so-called 'drainage networks' and pictures of Phobos and Deimos, the two Martian Moons.

The Mars 3 Orbiter/Lander sent up by the USSR in 1971 collected data for almost eight months. The Lander did land, but was more of a blink-and-its-gone story, gathering just 20

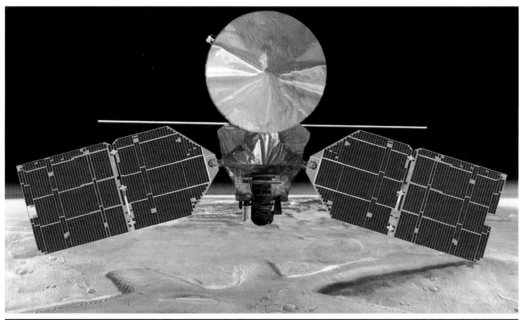

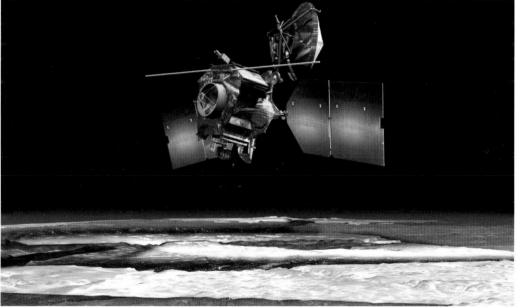

The Mars Reconnaissance Orbiter. Launched in 2005 by the US, this satellite is orbiting Mars with a bevy of cameras and radars to study minerals and water ice over Mars.

Credit: NASA

seconds of data before it packed up. The US added another first to its kitty with the Viking 1 Orbiter/Lander in 1975, the first successful landing mission to Mars. While its focus was to answer the 'is there life on Mars' question, and it conducted some experiments as part of this search, there was no concrete evidence that came back. It found that besides CO_2, Martian atmosphere also had nitrogen, argon, a mere 0.13 per cent of oxygen, carbon monoxide, water vapour in a minute 0.03 per cent and some inert gases. After this, there were quite a few Lander missions – Viking 2, again in 1975 itself, the Mars Pathfinder in 1996, Mars Explorer in 2003 and the Phoenix Mars Lander in 2007 – these were successful for the US. Obviously, with all this success, the focus shifted to rovers. On July 4, 1997, the Mars Sojourner travelled some 100 metres after landing on Martian soil, after having set off in 1996 as part of the Pathfinder mission. In 2003–04, the Mars Exploration Rover Spirit and Opportunity launches landed on opposite sides of Mars suitable for geological applications. Opportunity collected rock samples with past signs of water, and also spotted a mineral called jarosite that is formed only when there is acidic water available. There was also a Mars Global Surveyor launched by NASA, in 1997. There was the suggestion through its studies that liquid water perhaps existed just below

Wu Ji, Director General of the National Space Science Centre in Beijing and a key scientist for China's space exploration. India and China are regional rivals in the race for space, although not always admittedly so.

Credit: Pallava Bagla

the surface, but only in stray parts of the planet, also that the Martian poles would be good places to search for life, since they had water ice. On a Delta II rocket, taking off from Cape Canaveral in Florida, the Mars Odyssey mission took off in April 2001 and reached Mars in October of the same year. It had on it a payload meant to capture details of Martian climate and geological history. An elliptical orbiter, Mars Express (MEX) was launched by ESA on June 2, 2003, from Baikonur in Kazakhstan. It showed up larger quantities of water ice at the poles and also at the mid-latitudes, as did the Mars Reconnaissance Orbiter of the US. In a first of sorts, the US launched Phoenix as part of NASA's Scout Program, meant for smaller, lower cost spacecraft. It reached Mars in 2008 and brought back information of the alkalinity of Martian soil, with magnesium, sodium, potassium and chlorine. As for China, the Asian giant whom India likes to race against, although not always admittedly so, 2011 brought a Martian disappointment in the form of the Yinghuo-1, a Mars orbiter that was riding on the unsuccessful Russian Phobos-Grunt mission. Not that it deterred the Chinese in any way, for the country was already in the process of stream-lining its strategic planning for space activities and earlier the same year, had set up the National Space Science Centre to take overall charge of its space programmes under the umbrella of the Chinese Academy of Sciences in Beijing. Interestingly, in 2011 itself, China unveiled an ambitious five-year plan for space exploration that includes a human space flight programme and vast expansion of its Global Positioning System that encompasses both civilian and military applications.

All of this is obviously a lot of action for a planet so far away and so hostile in more ways than one, not to mention the exorbitant costs involved in undertaking deep space travel. That is why collaboration rather than competition seems to make sense, but then there are other factors to consider. In the long years of covering space

Indian **EXPRESS**

May 11, 1999

Destination Moon is ISRO's next big thing

PALLAVA BAGLA
NEW DELHI, MAY 11

TODAY may have been the anniversary of Pokharan II but what caused more excitement in the scientific community was chairman K Kasturirangan's announcement that ISRO's Polar Satellite Launch Vehicle can "undertake a mission to the moon." And a core team of scientists is being put together to work out the details.

In his Technology Day lecture here this evening on "The Indian Space Odyssey," Kasturirangan said that India could easily launch a small satellite of about 275 kg in a "fly-by mission" to the moon or even place a 140-kg satellite in an orbit around the moon. The mission: to study the moon's core. A manned mission, however, is still far away.

Destination Moon, he said, could symbolise the next big challenge for ISRO which has satellite technology well under its belt.

Working out the mission's objectives and payload could take time and if all goes well, it could be a reality by 2008.

The launch vehicle "will not be the problem" he said. The cost will be estimated, he said, once the scientific details have been worked out and the government will be approached for funds, Kasturirangan said. According to ISRO's plan, the Indian way to go to the moon could be by injecting a satellite that has a lot of onboard fuel into a lunar transfer orbit and then using onboard rockets to nudge the satellite closer to the moon. Later, there could also be a moon landing.

The May 12, 1999 front page Indian Express story which announced '*Destination Moon is ISRO's Next Big Thing*'. The first time the world heard that India was aiming for the Moon

The hot potato. American space scientist, Stewart Nozette, seen here in happier times when he was a collaborator on an American instrument that flew on Chandrayaan-1. Subsequently, he was arrested by the American authorities on charges of being an alleged spy for Israel. Despite this hiccup, India and the US continue to work together in space technology.

Credit: Pallava Bagla

General Charles F Bolden, NASA Administrator and former astronaut, who staunchly believes that the world's oldest and largest democracies can work together to pursue common goals and aspirations. He is looking forward to a joint Indo-US radar satellite mission.

Credit: Pallava Bagla

exploration, nuclear science and the politics of science and technology from India, but with an eye on the world, some big themes have always stood out for me, some incidents and events have become historic milestones, and some news has always pushed the frontier in one way or the other. These are those moments that make a reporter's life extraordinary. When I interviewed Bolden, the most natural opening question to ask him was to do with how he felt about the long hiatus in Indo-US space relations owing to the tight sanctions that came on India post the Pokharan blasts. I wanted to know what it felt like to deal with this long silence after the early, golden days. After all, the first steps of the Indian space programme were taken with full-fledged support from the US (the Nike Apache, an American rocket, was the first rocket launch from India back in the 1960s); even as late as 1982, India's first commercial communications satellite, the INSAT-1A was an American satellite built by Ford Aerospace and launched from the Cape Canaveral space station using an American Delta rocket. But India's Pokhran changed all that. In the wake of a developing nation's declaration of nuclear

Indo-US space diplomacy overtures around India's Mars mission. Seen in the picture is the then US Ambassador to India Nancy Powell visiting the launch pad ready for Mangalyaan, accompanied by MYS Prasad, Director of the Satish Dhawan Space Centre at Sriharikota.

Courtesy: US Embassy in New Delhi

capabilities came stringent sanctions from the US, heralding a long hiatus of estrangement on space exploration. Did this sadden Bolden, as the leader of the American space programme? Bolden's reply was sans any hesitation – 'That was history. What I think is most important is, what people should understand, is that in spite of the differences, the two countries have been able to come back together.' Indeed. Like the possibility of the development of an Indo-US Radarsat, a machine with day and night viewing capability.

Quietly, but assuredly, new relationships are blossoming too. India launched a spy satellite for Israel under the first term of the United Progressive Alliance led by the Congress Party, which was strongly opposed by the Left Parties. More recently, India bought a Radarsat from Israel. The small but high-tech country, also called a start-up nation, has openly spoken of collaboration. In Tel Aviv in Israel, attending the Ninth Ilan Ramon Annual International Space Conference in January 2014, one stumbles upon the country's ambitious plan to do

The Asian space race, with India and China leading the pack. This is a gathering of Chinese scientists who study the Moon and the Solar System. Seen here at an international conference in Udaipur in 2004.

Credit: Pallava Bagla

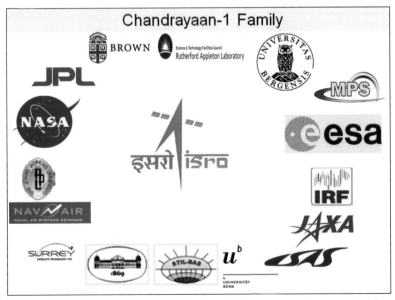

India was the captain of a team of 15 international partners who got together to undertake an exploration of the Moon on Chandrayaan-1 in 2008. This included big daddies like NASA and ESA.

Credit: ISRO

what only the world's biggest countries have so far managed to do – landing an unmanned spacecraft on the Moon as early as 2016. Till date, only Russia, the US and China have soft-landed on the Moon, and India hard-landed its tricolour using the Moon impact probe in 2008 sent there along with Chandrayaan-1. The washing machine-sized spacecraft that weighs 121 kgs is being readied by a not-for-profit venture called SpaceIL which operates out of a small nondescript office donated to it by Tel Aviv University. The Israeli lunar probe had its genesis after the USD 30 million Google Lunar X Prize was announced as a competition which challenged non-state owned space agencies to land on the Moon, send back photos, and move 500 metres on the surface of the Moon. About two dozen global teams are racing to win the prize and SpaceIL reckons it is in the pole position. A young ex-NASA engineer Yonatan Winetraub and two of his friends conceived of the spacecraft in 2010 sitting in a bar in Israel and then used a Facebook page to promote

Can India and Israel work together in space in the future? Seen here is the upper stage of the Israeli Shavit rocket. On rare display in Tel Aviv.
Credit: Pallava Bagla

the dream. Today, the dream has matured into a USD 36 million mission with 20 full time employees and 250 volunteers. Mr Winetraub said, 'Our aim is to put the Israeli flag on the Moon and to ignite the minds of youngsters.' He adds that if they win the USD 20 million prize, they will donate the money for children's education. Israel's highly developed space industry has backed the project by giving it support, both through donations and in kind. The country's President Shimon Peres is the first volunteer for the proposed mission, and Israel hopes to be a world leader in space technologies. 'It is a wonderful effort,' exclaims Menachem Kidron, Director General of the Israel Space Agency, adding, 'A great bunch of students are doing a good job to try and see if Israel can be the fourth country in the world to land on the Moon.' Around 40,000 school students have been associated with this project writ with 'national pride.' The yet to be named spacecraft will, among other things, take a 'selfie' of itself on the lunar surface and is also likely to carry out the first-ever seed germination experiment on the surface of the Moon under microgravity conditions. Mr Winetraub says Israeli school children will be asked to simultaneously replicate the experiment on Earth and the results will be compared. The young team is currently looking for the right rocket launcher to piggyback

Tiny Israel sets its sights to land on the Moon. Called SpaceIL, this is a private Israeli venture which is competing for the Google Lunar X Prize. Israeli engineer Yonatan Winetraub has sought ISRO's help to launch his country's maiden lunar dream.

Credit: Pallava Bagla

this Israeli dream. The team from SpaceIL is seriously considering hitching their Moon lander on India's Chandryaan-2 satellite that is likely to be launched in 2016, if they are able to negotiate a good deal with ISRO. Indeed the 'start-up' nation, as Israel is often described, with its seven million population is dreaming big of reaching the Moon in such a short time. India and Israel could be a marriage of endless possibilities in the twenty-first century, two countries born in the same year, and effectively using space for defensive security reasons. Tiny Israel sits in a hostile setting and one sees how small companies aided by the government are making small satellites with some of the best capabilities. These satellites, such as the Ofeq series, weigh a few hundred kilograms, have a life of several years and carry a single, high resolution camera. India also makes high resolution satellites in the Cartosat series, of less than one metre resolution. Israel is very keen to cooperate with India since the launch of Tecsar, an Israeli satellite, by the PSLV and the Indian space programme. And the subsequent acquisition by India of a similar satellite for day and night viewing capabilities. Menachem Kidron expressed his keen desire to cooperate with India saying the two countries have a similar hostile neighbourhood, intense defence cooperation, and in space both can be natural partners.

Now, for the first time in decades, NASA and ESA are planning the very first joint mission to Mars—the ExoMars Trace Gas Orbiter, set for launch in 2016, and expected to be the first in a series of joint robotic missions to Mars, whose major objectives are to study the chemistry of its atmosphere, with a 1,000-fold increase in sensitivity over previous Mars orbiters, to focus on trace gases, including methane. Meanwhile, NASA is not allowed to work with China in a

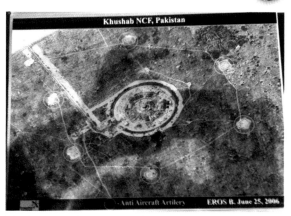

Tracking nuclear rogues from out there. This image of Pakistan's nuclear reactor complex at Khushab was taken by a high precision Israeli satellite. Seen around the reactors along the perimeter of the complex are fortifications in the form of anti-aircraft artillery positions.

Credit: ImageSat International

bilateral fashion. In the early days of the conceptualisation of the International Space Station (ISS), India missed the chance to be on this prestigious, international project. In January, 2014, ISRO shared the news that India had finally been invited to partner in the gigantic global effort that is the close to 4 lakh kgs flying machine the size of two football fields. ISRO Chairman Radhakrishnan confirmed that India would initially be contributing instruments with no immediate plans to send Indians to the space station. India will be the sixth nation to join this USD 100 billion effort. In 2013, ISS partners took the decision to keep the platform active and operational till 2020 at least, making an offer to space-faring nations to join up.

So, in this modern milieu, there is hardly any doubt that the future belongs to those who will collaborate, in small or large measure. ISRO and NASA are beginning a collaboration focused on asteroids. NASA's future Mars plans are ambitious. Sample returns, surface explorers, aeroplanes (that could cover more territory of investigation than rovers) and balloons, the greater use of precision landing techniques, and astrobiology field laboratory missions that would be expected to look for signs of life, past or present. Will India come to be regarded as one of the leaders at this frontier of cutting edge space science? As Bolden said, 'It's always exciting to have as many countries as possible participating in exploration efforts, particularly Mars. We were there with Curiosity, and we were able to carry along five other nations with us. It is exciting to have the United States and India join together and now getting ready to do more studies on the Mars's atmosphere with MAVEN (Mars Atmosphere and Volatile

Small country, big dreams. Israel is a powerhouse of frugal space technology. India bought what is seen here, the TecSAR satellite, from Israel and this spy satellite gave India the capability of keeping track of the activities of its hostile neighbours. India also launched the same satellite for Israel.

Credit: Pallava Bagla

Evolution) looking at the upper atmosphere of Mars, a place that we don't know a lot about. The Indian mission is also looking at the Martian atmosphere.' To Radhakrishnan, India is already in a leadership position – 'Certainly the Mars mission is going to create a great sense of pride, a great sense of achievement and it is going to also improve our position in the international community. And tomorrow, when there is a large mission by the global community India is going to be there as one of the partners.'

NOTES

1. P. Bagla, Indo-US Space Ties Ready for Take-off: NASA Chief. Interview with Administrator of NASA, General Charles Bolden, *The Hindu*, August 29, 2013.

2. '12th Five Year Plan of the Department of Space', Report of the Working Group (WG 14) Web Version, October 2011.

3. Ibid.

4. U.R. Rao, *India's Rise as a Space Power*. Foundation Books, Cambridge University Press India Pvt Ltd, 2014.

5. Mars Orbiter Mission Study Report, ISRO, July 2011.

Chapter 8

The Years Ahead: The Future of India's Space Programme

In the young, a community sees hope, and a nation sees its future. Half of India's 1.25 billion people are younger than 25 years of age, and 65 percent of the country's population is under the age of 35! This is a stunning demographic that could be the world's envy, but not unless the country does enough for its young. That is perhaps why any leader with wisdom would ensure that Indian youth be acknowledged, nurtured and supported as much as possible. Their needs, aspirations, dreams and ambitions cannot be met without specific strategies and agendas of transformation. No wonder PM Modi, when addressing space scientists at Sriharikota said: 'Development of human resources will be critical for our future success. I was very pleased to meet our young scientists here. I admire their work and their achievements. Let us link up with more universities and colleges, to develop our future leaders in this area. We must also involve our youth at large, with space. Let us use social media to further engage with our youth. Let us invite school and college children, to witness launches and visit space centres. Could we also think of developing a state-of-the-art, interactive, digital space museum?' Undeniably, the value of a high calibre, even spectacular museum as a learning and exposure tool would be a great contribution to space science and technology in India. Unfortunately, the art and science of creating and sustaining museums has eluded most Indians, and whatever the thematic area, museums in India are either shabby, ill-kept or terribly designed, unattractive and completely ineffective. This is particularly surprising because India is a land of diverse art forms and creative arts and crafts, of several institutions of higher learning in design, and of extraordinary

variety in these forms that differ from state to state, region to region. So, maybe, ISRO ought to take up PM Modi's challenge and work with the best artistes and designers and communicators in the country to make this museum a reality. Resources, political support and commitment are all on ISRO's side, but there is little doubt that the organisation needs to push its public engagement to the next level.

The future, after all, is filled with excitement, some planned, some blue sky. 'Why not visit Mercury?' is the question UR Rao, former Chairman of ISRO and now Chair of the Governing Body of the Physics Research Laboratory in Ahmedabad, asks, sometimes of himself, sometimes openly. He speaks of this possibility somewhat distractedly as I sit across his table at the ISRO Headquarters in Bangalore. Well, why not, I wonder? Although Rao speaks more in the context of how this query bounced in when the Advisory Committee on Space Sciences (ADCOS) that he heads was wrestling with the Mars mission proposal, it seemed to be a perfectly valid challenge for the space folk. Rao goes on to elaborate the point, explaining that he would like to see India venturing into those unexplored corners of space that other nations have not tried out as yet. As a matter of fact, that would be the real thing when it comes to breaking new ground. Otherwise, even today, India tends to go where others have gone which in itself is hardly anything to quibble over, if one is to consider the rather tall achievements of the space programme, but does deny the country the allure of breaking truly new ground that is at the centre of scientific discovery.

But who's complaining? ISRO's calendars for the next five years are action-packed. The fine-mannered and gentle Dr S Unnikrishnan Nair, on whose somewhat slight shoulders sits the responsibility for planning the realisation of the Human Spaceflight Programme, may have his constraints in being able to share much detail about what is really a pre-project activity at VSSC in Thiruvananthapuram. But, gazing at a bright orange prototype of an indigenously-developed astronaut's suit, one must say the anticipation does build up. Indians in space would be such an exciting and significant milestone for the country. Unnikrishnan, the great lover of plants, is not at all uncomfortable indicating that there is not much he can say at this stage! But what he does share is the action around some of the critical technologies that are being developed around the HSP including the fast-paced work on testing a real crew module in a rocket. Needless to say, sending an Indian into space from Indian soil using Indian rockets is no longer just a pipe dream and who knows, a self-confessed space buff like PM Modi may just propel it into reality.

The Sun is the king of the solar system. Early morning Indian cityscapes of people standing with folded hands outside their homes, in parks, on their porches and verandas, bowing their heads to the rising Sun, show this reverence. Ancient Hindu Vedic texts are filled with literary odes to the Sun. Some are devoted just to the Sun, like the *Surya Upanishad*, an ancient Indian Hindu text that is part of the *Atharvaveda*, or the *Rig Veda* in which this elegant *shloka* gathers all that the Sun means to life in three words – '*Surya Atma Jagatastasthushashcha*' – which means that the Sun God is the soul of all beings, moving and non-moving. No wonder the Sun is worshipped across diverse religions and cultures in the world, and the Hindus call it *Suryopasana*. In the *Ramayana* too is the well-known *Aditya Hridayam* (which literally means 'heart of the Sun'), part of the narrative of Valmiki's *Ramayana* where the sage Agastya helps Rama begin a prayer to the Sun to strengthen and uplift his sagging spirits when he is facing Ravana on the battlefield. Aditya is one of the many Sanskrit names by which the Sun is known, and just like most of the synonyms of the Sun, is a popular name for boys in India. For space buffs, the Sun is the great centre of the solar system so when ISRO announced its intent to mount a mission to the Sun in the near future, nobody was surprised. Except, of course, to wonder how anything manmade could withstand its blazing temperatures and raging solar winds! And mirroring the ancient texts, the mission has been aptly christened Aditya, which will be an Indian spacecraft setting out to explore the Sun from the Earth's orbit.

Aditya-1 was born, as an idea, in 2008 and G Madhavan Nair, ISRO's Chairman at the time, didn't waste much time in sharing the good news, which he did in November the same year, announcing a scientific mission designed to study the Sun's corona, which is the outermost region of the Sun. The corona is made of hot, ionised gas and the temperature out there is obviously unthinkably hot, at about two million degrees Centigrade! So, what is it that heats up the solar corona and what gives the solar winds their acceleration—these are some of the major scientific objectives of the proposed mission in which a space solar coronagraph will be the main instrument to study the Sun and get a sense of the physical processes out there. The plan is to launch Aditya-1 with five scientific instruments to be parked in an orbit 1.5 million km above the Earth. The coronagraph will provide high quality images that scientists would be able to study.

These are thrilling times at ISRO. Flying into the face of the Sun, arranging for Indians to set off on a manned mission into space, or preparing to land on the Moon with a rover – future plans look bright. With the Indian space programme having consolidated its major pillars of activity and programmes, perfected primary space exploration studies and having become almost fully self-reliant across these domains in the last fifty years, ISRO seems to be on a roll. That is why the next few years, up until 2020, are designed to push ISRO's productivity to more exciting realms by getting its staff and laboratories and institutes to work on 'advanced launch vehicle systems, thematic Earth observational satellites with improved resolution, high power throughput communications satellites, microwave, multi-spectral remote sensing

The Moon Impact Probe which carried the country's tricolour onto the surface of the Moon. In the Asian space race India undoubtedly sprinted ahead of China in staking its claim on the Moon.
Credit: ISRO

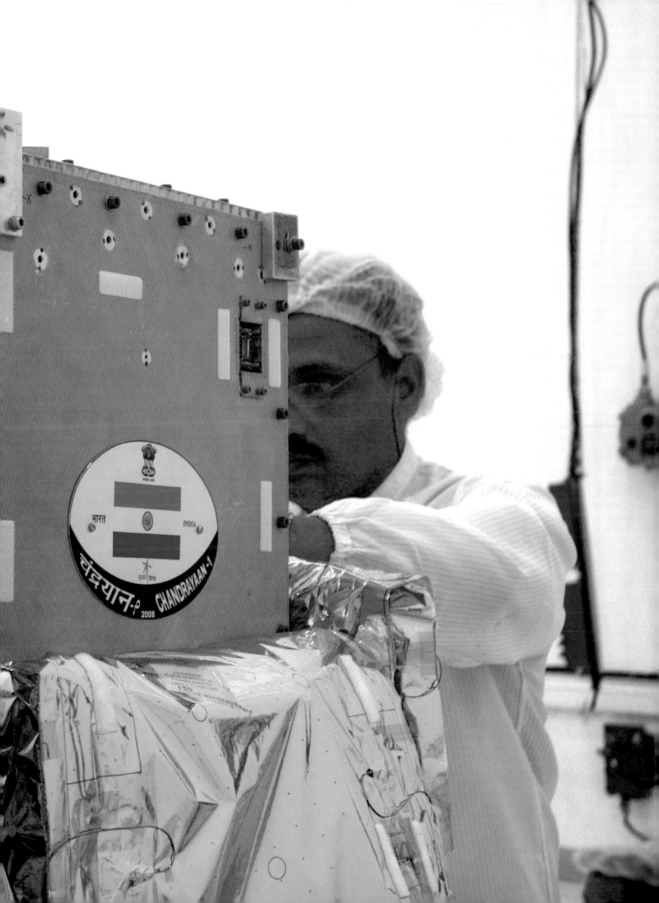

satellites, weather and climate studies, a constellation of satellites for regional navigation, development of critical technologies for human spaceflight and satellites for space science and planetary exploration purposes.'[1] A total of 58 missions are projected to be undertaken during the 12th Five Year Plan (2012–2017) which includes 33 satellite missions and 25 launch vehicle missions.[2] And what's more, the societal applications of these activities seem to be growing too. A few years ago, former PM Singh was addressing a gathering at SAC in Ahmedabad, where he gave a rousing, motivational speech: 'We have proved all those wrong who claimed that a space programme is a luxury that developing nations cannot afford. Our space programme has, in fact, helped us to leapfrog in technology and bring significant social, economic and industrial transformation to the most remote areas. With relatively modest financial outlays, we have put in place a space infrastructure that touches every aspect of an ordinary citizen's life. It has reduced uncertainties and ignorance, and shrunk time and distance. It has brought Indians closer to each other than we have ever been.'

Although things are a bit hush-hush on the Human Spaceflight Programme (HSP), with government approvals and permissions still awaited, Dr S Unnikrishnan Nair who heads the HSP from VSSC is busy, along with his team, focusing on the several pre-project activities that are part of the run-up, indispensable as they are for a manned mission, as and when it happens. Nair explains how his team's current mandate is to 'develop critical technologies that are required to be in place before starting the project.' Doubtless, the HSP, as and when it rolls out, will create huge excitement across the country. India has had the privilege of its citizens travelling into space, but never from Indian soil on Indian missions. Wing Commander Rakesh Sharma was with the Indian Air Force and flew as an astronaut on the Soviets' Soyuz T-11 in 1984 making history as being the first Indian to visit space. One of the indelible memories that he created was during the conversation he had with the then Prime Minister Indira Gandhi while he was still in space. She asked candidly, with a smile on her generally stern face: '*Upar se Bharat kaisa dikhta hai aapko?*', meaning 'how does India look to you from up there?' Sharma answered without a moment's hesitation: '*Saare Jahan se Accha,*' meaning 'nicer than the entire world.' Naturally, Sharma remembers the moment clearly even today, and chuckles about his desire then to give a snappy, happy response to Mrs Gandhi. He probably took his inspiration from the famous Muhammad Iqbal, one of the best known Urdu poets and philosophers who lived when the British ruled India, by borrowing the opening line of a famous patriotic song written by Iqbal. Interestingly, Iqbal happened to be one of the key people who inspired the movement for a separate state for Pakistan, but the iconic poem he wrote is today one of free India's most popular patriotic songs, sung at almost every occasion that is related to celebrating freedom, nationalism and pride in being Indians. There must be many Indians who would like to emulate Sharma's achievement. Adding to such inspiration

India's second mission to the Moon, Chandrayaan-2 hopes to land a rover on the lunar surface. Seen here is an early iteration of the rover.

Credit: ISRO

are other world-renowned astronauts of Indian origin, both women, interestingly. Dr Kalpana Chawla was working with NASA and unfortunately made greater news in her tragic death, when the space shuttle Columbia met with disaster in 2003 after re-entry into the Earth's atmosphere during its return. She had flown into space once before, in 1997. Chawla lived and studied in Karnal, Haryana, and Chandigarh, before she went off to the US to pursue higher studies. Sunita Williams, a pilot with the US Navy, is another astronaut with Indian ancestry, and holds several space records to her credit. Williams was assigned to the International Space Station a number of times and is said to have accomplished the maximum number of space walks by any woman, the longest space flight by any woman, and clocked the maximum spacewalk hours by a woman. Both Chawla and Williams, owing to the fact that they were women with Indian roots and were women who were astronauts, and because Williams visited India in 2007, catalyse a lot of excitement among Indians. And this is not just because they seem like they do exciting things, but because there are millions of children in India who dream of becoming astronauts. Whenever I speak to young children about the fascination of

Wing Commander Rakesh Sharma (extreme left) made history when on April 2, 1984 he blasted into space on a Soviet rocket. Also in the picture are the Spaceship Commander, YV Malyshev and Flight Engineer GM Strekalov.

Credit: Nehru Planetarium, New Delhi

The original space capsule in which cosmonaut Rakesh Sharma flew into space, also seen is the original space suit that was to be used by Ravish Malhotra.

Credit: Nehru Planetarium, New Delhi

A letter from a school student

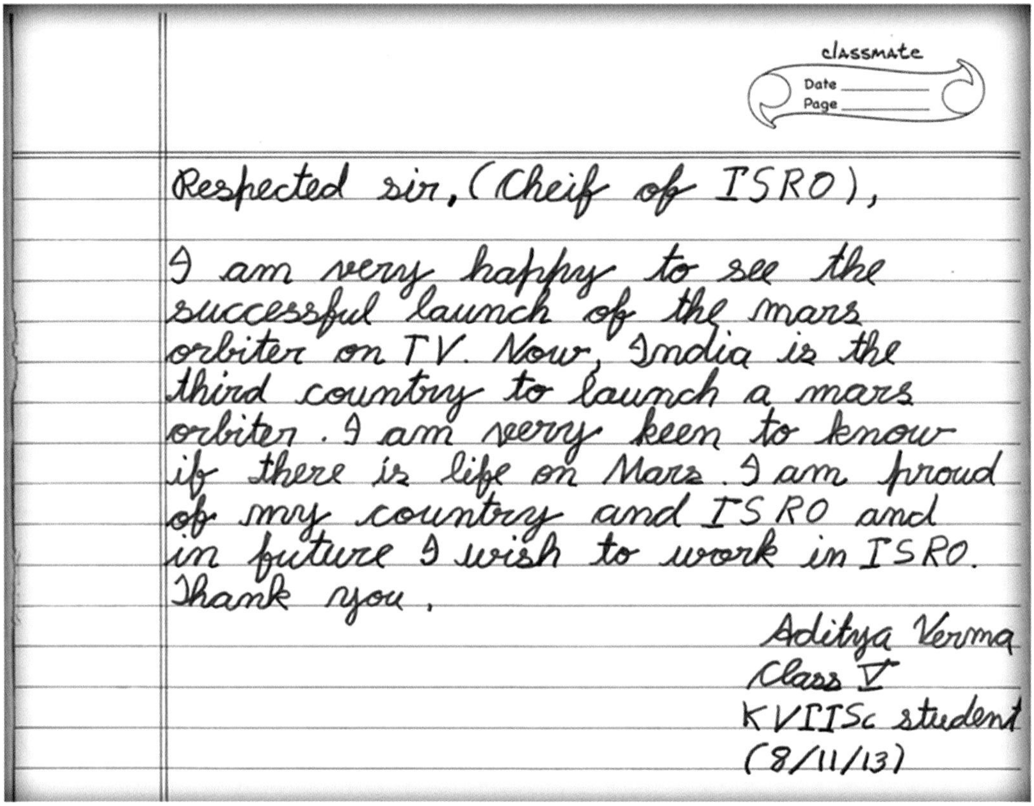

classmate
Date _____
Page _____

Respected sir, (Cheif of ISRO),

I am very happy to see the successful launch of the mars orbiter on TV. Now, India is the third country to launch a mars orbiter. I am very keen to know if there is life on Mars. I am proud of my country and ISRO and in future I wish to work in ISRO. Thank you.

Aditya Verma
Class V
KVIISc student
(8/11/13)

A ten year-old writes a letter to the ISRO chief, asking him if there is life on Mars. Inspired by the Mars mission, this child expresses a wish to work with ISRO.
Courtesy: ISRO

space exploration and the Indian space programme, almost every child shares his or her dream of becoming an astronaut someday. General Bolden said to me that he thought every Indian child should aspire to become an astronaut, quite like American children. That is why the current Indian plan to mount a mission in which Indians will lift off into space, most likely spend about a week out there, and then come back, is really significant because it will be a major first milestone. Sharma explains that the real value would come from building streams of study in such a way that young people actually become career astronauts, and that the link with military aviation may actually be essential in pursuing such interests.

1984 saw the first Indian astronaut go into space. Two Indian Air Force personnel were carefully selected for this mission, Rakesh Sharma (L) and Ravish Malhotra.

Credit: Nehru Planetarium, New Delhi

Kalpana Chawla. Indian-born American astronaut from the small town of Karnal in Haryana who made tragic history by her death when the American space shuttle Columbia disintegrated on February 1, 2003 as it entered the Earth's atmosphere on its way back from space, killing all seven crew members.

Credit: NASA

Among the many important pre-project activities that are happening with regard to the HSP, the crew module is being readied, tested and being given final shape, as is the crew escape system (see Annexure 8). Nair says: 'In such an undertaking, it is the safety of the crew, rather than the success of the mission that is most important to us.' Some have questioned the necessity of such a mission, and since it is expected to cost around ₹ 125,000 million, the questions are but natural. How a human space programme would bring benefits as compared to satellites in space is leading critics to wonder whether such programmes are meant only for India to show its prowess on the world stage. Anyway, the decision to initiate work on the HSP came through several national consultations and final Space Commission discussions with Mr Montek Singh Ahluwalia, former Deputy Chairman of the Planning Commission, back in February, 2009. ISRO's plan to put humans into space got in-principle approval with funds allocation to be distributed through the eleventh and twelfth Plans, and work began in 2009 itself. But, over the last few years, it seems the full throttle action on the HSP has slipped in priority. 'A decision that involves more than ₹ 100,000 million surely needs necessary consultations,' says a calm Nair, least perturbed that the actual decision to get into mission mode on the manned spaceflight programme is yet to come. Another reason for the tardiness in taking a decision has also been the setback that the Indian space programme suffered when the GSLV as a functional, operational vehicle met with twin disasters in 2010 and took time to recover. ISRO's assessment is that it will take roughly seven years from the time full approvals are received to when Indian astronauts can actually go up into space. Since the government has just given clearance for a study project, the engineers are somewhat tight-lipped on details, and the HSP has not received much publicity. As it is, working on man-rated systems for space travel is a completely different ball game from what ISRO has done so far, and therein lies the challenge.

None of this seems to take away from the significance of events and activities that are happening in the run-up. The bright orange space suit that has been created as a prototype with the Indian flag prominently displayed makes it easy for the mind's eye to imagine Indian astronauts up in space. Using the ₹ 1,450 million sanctioned by the government for the development phase, milestones are being reached, bit by small bit. The crew module, for instance, is to fly as a dummy spacecraft on the first flight of the GSLV Mk III. It is the GSLV Mk III that seems to be the appropriate launch vehicle that would need to be perfected for being worthy of a human spaceflight mission. The idea is to test out the module, of which the outer skeleton was manufactured by HAL and delivered to VSSC in February 2014, by keeping an eye on

Sunita Williams. An American astronaut of Gujarati origin, she carried the Bhagavad Gita, an idol of Ganesha, and samosas into space.

Credit: NASA

Frugal space missions make India an attractive destination. A team from the European Space Agency visits ISRO's Satellite Centre.

Credit: Pallava Bagla

Bird's eye view of India's power centre, Lutyen's New Delhi. One can see the entire central vista from Rashtrapati Bhavan, the regal home of the President of India, atop Raisina Hill, going down to Vijay Chowk where one can literally count each car that is on the road, and further down to the iconic India Gate and the legendary National Stadium, home to India's hockey team. This is an image taken by an

ISRO remote sensing satellite. Today, ISRO has the capability to map objects less than a metre in size from its birds in the sky being flown at 800 to 900 kms above the Earth.

Credit: ISRO

how smoothly the re-entry process and flight dynamics are able to operate. Equipped with systems necessary for crew support, navigation, guidance and control, the module, when finally used for its core purpose, is to be recovered some 400-500 kilometres off Port Blair in the Bay of Bengal.[3] Work is also ongoing to refine and perfect the flight suit and the crew escape system, and develop the Environmental Control and Life Support System (ECLSS). There are other things to be done, such as parachute developments and working with the Institute of Aerospace Medicine in Bangalore to get a fix on the medical and health-related requirements of astronauts in space. Nair also says that a lot of support is being sought from industry, such as with fire safety technologies. Chairman Radhakrishnan has often said at various platforms that these are just the early steps that ISRO is taking on manned missions, and that this phase when critical technologies must be developed is important because, as he says 'these are things we have not done in the past'.[4] Finally, a fully autonomous orbital vehicle will carry two or three crew members to about 300–400 kms above the Earth into the LEO

The US may have abandoned its space shuttle but India dreams of making a reusable launch vehicle which would reduce the cost of space transport.

Credit: ISRO

The future is here. ISRO recently tested an air-breathing engine as part of its Advanced Technology Vehicle (ATV) project. This is meant for substantially reducing the cost of space transportation.
Credit: ISRO

and bring them back safely, to a pre-defined destination on Earth.[5] The idea is to eventually build and demonstrate India's capability to undertake manned missions. Till date only Russia, the US and China have successfully flown astronauts into space on their own indigenous systems, with China having logged the achievement in 2003. Overall, some ₹1,450 million is being used to develop critical technologies for the HSP. Nair says that 'our core team is small, we interact with various entities at ISRO, and a lot of technology development is happening, that is our mandate right now.' Going by ISRO's work tradition, it is quite likely that all these efforts will jog along at a decent pace till the final green signals flash, after which the entire organisation will throw its might behind the project, as it has been with the Mars mission. This style of working allows the organisation the leeway to get inception activities going and ready the foundation for a large project well in time to avoid last minute scurries. There are offers too, the US and NASA have often said that the country would be willing to train astronauts for India at its facilities, but of course the costs attached are bound to be prohibitive.

Rocket of the future. The Geosynchronous Satellite Launch Vehicle (GSLV)—Mark III. ISRO hopes that this would be the preferred vehicle for launching Indians into space on an Indian rocket from Indian soil.

Credit: ISRO

Two happy astronauts. Sunita Williams, the US astronaut of Indian origin along with Rakesh Sharma, India's home-grown astronaut.

Credit: Pallava Bagla

Meanwhile, a natty-looking Rover is getting perfected elsewhere in ISRO, crawling around on an artificial surface created to simulate what it is like on the Moon, at I-SITE in Bangalore. Following the success of India's first Moon mission, Chandrayaan-2, in an advanced *avatar* of the earlier mission, is on its way. Initiated in 2008 as a joint mission between India and Russia for *in-situ* studies of the lunar surface, Chandrayaan-2 underwent programmatic re-alignment during 2010–12. As Dr K Kasturirangan, former Chairman of ISRO said: 'It is not a question of whether we can afford to go to the Moon. It is whether we can afford to ignore it'.[6] This time, India hopes to test its capability to soft-land on the lunar surface with a lander carrying the home-made Rover. If this is successful, ISRO's Moon studies will surely move to the next level. Chandrayaan-2 will have a two module system with one Orbiter Craft module (OC) and a Lander Craft module (LC) carrying the Rover. This exciting piece of machinery is likely to be designed as a 6-wheeled structure, and is expected to be released by the Lander Craft. With power drawn from a small solar panel, the Rover is envisaged in such a way that it will perform mobility activities in the low gravity and vacuum of the surface

Preparing for a human space flight. ISRO in the midst of perfecting critical technologies necessary for a human space flight, which it estimates would cost ₹ 125,000 million.

Credit: Pallava Bagla

of the Moon using its semi-autonomous navigation and hazard avoidance capabilities. The Rover's communication links will be with the IDSN at Byalalu near Bangalore, either through a Lander Rover Communication System on-board the Lander, or through the Orbiter Rover Communication System on-board the Orbiter. The Orbiter Craft, as its name suggests, will go round remote sensing the Moon, along with its payloads meant for scientific inquiry conducting mineralogical and elemental studies of the surface of the Moon. As of now, the recommended payloads the orbiter will carry are five, and there are two meant for the Rover.[7] Dr Anil Bhardwaj at the VSSC is excited about working on one of the payloads called ChACE 2, similar to ChACE-1 on Chandrayaan-1. This instrument will study the lunar exosphere in detail, and is basically a spectrometer. ChACE-1 was the instrument that found water on the Moon before the US instruments on-board Chandrayaan-1, but since it was not calibrated properly, the true credit for this discovery went to M^3, also flown then. So, what could have been a fully Indian discovery became an Indian-American one! Yet, none of this has detracted from the significance of Chandrayaan-1 having located water on the Moon, and the landmark discovery does belong to India since the country was the captain of the ship, so to speak. Aboard Chandrayaan-2 there are other payloads envisaged, a Large Area Soft X-ray Spectrometer (CLASS) and a Solar X-ray Monitor (XSM) for mapping major elements of the lunar surface; the L and S band Synthetic Aperture Radar (SAR) for probing the first few tens of metres of the lunar surface for the presence of different constituents including water ice; an Imaging IR Spectrometer (IIRS) to map the lunar surface over a wide wavelength range for the study of minerals, water molecules and hydroxyl ions; and a Terrain Mapping Camera-2 (TMC-2) that will prepare a three-dimensional map

Dr. K. Kasturirangan, former Chairman, ISRO, under whose leadership the Moon mission was conceived and announced on May 11, 1999.

Credit: Pallava Bagla

that would be invaluable in studying lunar mineralogy and geology. On the Rover, there would be a Laser Induced Breakdown Spectroscope (LIBS) and the Alpha Particle X-Ray Spectrometer (APXS) to conduct elemental analysis of the lunar surface near the landing site.[8] The Lander will also contain a probe which will go to the surface of the Moon and capture temperature profiles at different places. Chandrayaan-2 has also been delayed because of the GSLV's progress.

The development, fabrication and launch of an astronomical observatory for studies of cosmic sources is also on the cards, through a forthcoming satellite called the ASTROSAT project. ASTROSAT is envisaged to be a national observatory available to any researcher in India for astronomical observations, and will be a multi-wavelength space-borne observatory that would enable simultaneous observation of the celestial bodies in ultra-violet, visible and x-ray bands (see Annexure 9). Although most of the observation time will be for the use of Indian researchers, a part of the ASTROSAT observation time will also be made available to the international astronomical community on a competitive basis. The plan is to launch ASTROSAT using the PSLV, with a configured lifetime of a minimum period of five years. In an interesting collaboration, the six scientific instruments onboard ASTROSAT have been developed by ISRO, the Tata Institute of Fundamental Research, Mumbai, and the Indian

Institute of Astrophysics, Bangalore. The Canadian Space Agency is also collaborating for the development of one of the scientific instruments.

India's very own GPS, designed to provide an accurate position information service to users in India as well as the region extending up to 1,500 kms from its boundary, the Indian Regional Navigation Satellite System (IRNSS), mentioned earlier in the book, is being realised as an independent regional navigation satellite system with a seven-satellite constellation by 2015. Two satellites in the IRNSS constellation are already operational in orbit and two more are expected to be launched during 2014. IRNSS satellites are small, identical in configuration and therefore have a smaller turn-around time. Locating accurate position, time and navigation will be a key service.

In the world of space exploration, the idea of reusing a launch vehicle is a powerful one, bringing with it cost effectiveness, a reduction in labour and several other advantages. The space shuttles that the US used were such vehicles and Columbia logged 28 successful flights between 1981 and 2003, when it met with the infamous disaster that cost the US the lives of

When space technology meets the ocean. Indian soldiers recover from the Bay of Bengal the Space Capsule which was made to orbit around the Earth in 2007.
Credit: ISRO

A prototype of the human space flight suit that is to be worn by astronauts who will fly into space, displayed at the Thumba museum.

Credit: ISRO

its entire crew at the time. The US space shuttle programme was retired in 2011. That said, the value of reusing launch vehicles remains a key point of interest for ISRO. It is therefore, hardly surprising that for the men and women at ISRO who focus on launch vehicles, nothing would be more exciting than taking on newer and newer challenges in the world of launch capabilities, where, quite literally, the sky is the limit. The rocket engineer Somanath, who passionately chases his dreams of creating better and better rockets, looks forward constantly to these challenges: 'We have concepts and designs available for the Reusable Launch Vehicle Technology Demonstrator (RLV-TD). It has to be taken ahead, which of course is not a small step, it requires ₹100,000 to 200,000 million. As for us, we are ready to try.' The RLV is a complex piece of engineering, designed to ensure that a launch vehicle can come back to Earth after it places a spacecraft in orbit. Flight models are currently being made on the ISRO shop floors and the RLV is certainly going to be a flagship technology that is expected in the near future. An Indian space shuttle, therefore, might just be a reality. Technology capability development is always a difficult nut to crack because industrial partnerships are hard to come by, given the risks. According to ISRO, the RLV-TD, already configured, is meant to become a test bed to help ISRO evaluate various technologies such as hypersonic flight, autonomous landing, powered cruise flight and hypersonic flight using air breathing propulsion, each a mini-challenge in itself! The RLV-TD will be involved in a series of technology demonstration missions over the next few years.

Trying to retrieve what you send out into the dark emptiness of space is indeed critical for the future of space programmes. This is another frontier that ISRO is trying to push, through the Space Capsule Recovery Experiment (SRE) that began in the mid-2000s and was focused on fine-tuning technology that allows for reclaiming a space capsule from somewhere in the Universe, something that can evidently have multiple advantages, not the smallest being its contribution to human spaceflight. SRE-1 was a 550 kgs capsule that was launched into a 635

The space capsule that came back. ISRO conducted a Space Capsule Recovery Experiment on January 10, 2007. This 555 kgs satellite was launched into an orbit 637 kms above the Earth, and then recovered in the Bay of Bengal. This was the first and most important step in dreaming of a human space programme.

Credit: Pallava Bagla

ISRO preparing to launch Indians into space. A prototype of the crew module which can accommodate two to three astronauts.

Credit: ISRO

km polar Sun-synchronous Orbit in January 2007 as a co-passenger with CARTOSAT-2. It remained in orbit for 10 days and during that time the payloads it carried performed operations they were intended to, after which the capsule was de-boosted and recovered successfully back on Earth on January 22, 2007, when it splashed down into the Bay of Bengal, and its retrieval was supported by the Indian Coast Guard and the Indian Navy. Understanding how re-entry happens, and how it can be guided and controlled was a key learning from this experiment, as was the development of technology that allows reusable thermal protection given the massive temperature ranges that space travel endures. As a follow-up, the SRE-2 project was conceptualised with the main objective of realising a fully recoverable capsule and to provide a platform to conduct micro-gravity experiments.

On a similar track of futuristic thinking is the idea of reducing the cost of accessing space, and ISRO's air breathing technology and scramjet engines based on the idea that using atmospheric oxygen as a catalyst for combustion of on-board fuel will be a great improvement over the current situation, in which conventional rockets are carrying not just the fuel but also oxygen on board. Air Breathing Propulsion (ABP) will obviously lead to a fairly significant reduction

Space in the service of the common man. Assets in space play a critical role in disaster management. Scientists of the National Remote Sensing Centre at Hyderabad analysing the tragedy that struck the temple town of Kedarnath in 2013.

Credit: Pallava Bagla

Caught in a storm: A skeleton of the GSAT-6, an ISRO satellite meant for providing seamless multimedia communication across India, is one of the fallouts of the alleged Antrix-Devas S-Band scam.
Credit: Pallava Bagla

in size and cost, besides improved vehicle operability. The scramjet engine with its supersonic combustion is being seen as the best and most efficient option for this kind of technology, and VSSC is working on Advanced Technology Vehicles (ATVs) that are new generation high performance sounding rockets and test flights have already happened over the last few years.

While all the new frontiers get explored one by one, taking ISRO to new heights and allowing a full-throttle space programme to take shape, the Indian population is growing at an alarming rate. Burgeoning with this is the demand for communications and other satellites and their services. It is a need that is almost like an unstoppable hunger. Bhuvan is an online geoportal that ISRO has created to ensure access to data and this web-based presence is growing too. Radhakrishnan reveals the ISRO plan, saying: 'During 2014–17, ten communication satellites are planned to be realised and orbited to provide 180 more transponders for in-orbit replacement of ageing satellites, and enhancement of national capacity, and enable introduction of new communication capabilities such as broad-band data connectivity and digital multimedia.'[9] So, it goes without saying that a large number of communications satellites

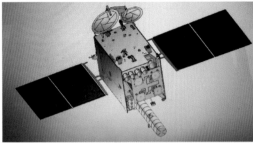

Defence always takes front seat: the GSAT-7 is a dedicated communications satellite for the Indian Navy and was launched on August 29, 2013. This was the first but will not be the last.

Credit: ISRO; Arianne Space

Non-Resident Indians have played a role in NASA. Alok Chatterjee from the Jet Propulsion Laboratory of NASA played a key role in Chandrayaan-1 and hopes to play a seminal role in the future Radarsat mission that is on the anvil, as a collaborative venture between India and the US.

Credit: Pallava Bagla

are on the agenda, many of them GSATs in a continuous stream, since communications requirements of people never reduce, but only grow. GSAT-6 is a high power S-band communication satellite that will be co-located with GSAT-12, INSAT-4A and GSAT-10 and will be so designed that it would be able to communicate with smaller ground terminals. This satellite will also provide a platform for developing technologies such as the demonstration of large and unfurlable antennae in satellites, hand-held ground terminals and network management techniques that could be useful for satellite-based mobile communication applications in the future. But GSAT-6 has had a troubled past, with its linkage to the Antrix-Devas scam, and the two are still slugging it out in a USD 2.5 billion arbitration. Following this would be GSAT-6A, expected to be launched by the end of the XII Five Year Plan, GSAT-9 and then GSAT-11. The latter is meant to be an advanced communications satellite. Then there would be GSAT-15, a communication satellite expected to function with a mission life of more than 12 years.

In partnership with the Jet Propulsion Laboratory of NASA, ISRO is working to create a dual frequency microwave imaging satellite, with India being responsible for the overall satellite platform, S band Synthetic Aperture RADAR and launching the satellite using the GSLV. The future is action-packed, there isn't much doubt there. Private space ventures are also finally taking shape, as with the Team Indus group in Bangalore, who

Satellites can be fun. Students from the Vellore Institute of Technology participating in a competition to make the best 'CanSAT', a nickname for a satellite that has to be made in such a way that it can fit into a fizz drink can.

Credit: Pallava Bagla

won a place to fight it out for a USD 40 million award Google Lunar X Prize for their Moonshot strategy, a lander with a rover meant to go off to the Moon in 2015. A group of young entrepreneurs led by Rahul Narayan, one of Team Indus' co-founders who calls himself

a serial entrepreneur is really living a childhood dream in trying to complete this fairly challenging project in time and to perfection. 'We have completed our design, and we are now building the craft,' says Narayan, who is very proud of his very young team of 20-somethings. Rushing towards a December 31, 2015 deadline given by Google Lunar X with what is as of now a small budget of USD 1.25 million, the group of young aerospace engineers and other professionals have big ambitions, hoping to walk their craft some 500 metres on the Moon. ISRO could be helping with the launch by providing the PSLV at a cost. Meanwhile, Mars One from Amersfoort in the Netherlands is aiming even higher, to set up a permanent human settlement on Mars from 2024 onwards. Having invited in 2013 applications from potential astronauts ready to make the trip, in May 2014, the Mars One group announced that just 705 potential astronauts would become the first human colonists on Mars sometime later this century. Some 20,000 Indians had registered for this one-way trip to Mars! Is this preposterous, is

A scientist from the Space Science Data Centre displays the very special fire and flood proof safe in which a back-up of the scientific data is stored.

Credit: Pallava Bagla

this freakish, or is this just inter-planetary adventure? That is one question with no answers right now.

For ISRO, growing the business is a focus for the future. Responding to a Parliament Question in 2014, Jitendra Singh, Minister of State (Independent Charge) for Science and Technology referred to how the Department of Space had plans chalked out till 2020, and also of how ISRO has gathered revenue to the tune of 39.82 million Euros from launching 15 foreign satellites, between 2011 and July 2014. A space park is likely to come up on Sriharikota or nearby, and would house all the industries catering to the Indian space programme. The plan is to carve out this 250 acre dedicated 'Space Park' adjoining the space port at Sriharikota which could be used by private players to develop the Indian aerospace industry. ISRO is also thinking of hiving off the money spinning communications satellite business, and launches are also becoming increasingly viable as commercial propositions. 'India has the potential, to be the launch service provider of the world. We must work towards this goal. Construct the required new launch infrastructure. And extend our launching capabilities to heavier satellites,' said PM Modi. ISRO naturally banks on growing the frequency of launches. That is why there is action at Sriharikota too, in terms of expansion. A third vehicle assembly building is already being constructed, and a third launch pad is also likely to come up as the HSP draws

The National Space Science Data Centre. Banks of computers keep safe data gathered through Indian space exploration missions.

Credit: Pallava Bagla

Future tense. Securing strategic assets of space on the ground would become a huge challenge in the future. Armed guards perched on a watch tower keep vigil over India's spaceport on the Bay of Bengal coast.

Credit: Pallava Bagla

closer, meant as a dedicated site for training astronauts. Two sites, one at Machlipatnam north of Sriharikota and another in Tamil Nadu, are also being examined for a second launch site for the Indian space programme. Another possible destination is the Andaman and Nicobar Islands, but the potential threat of a tsunami makes it less attractive. There is also the likelihood of acquiring an island on lease from Mauritius; but the easiest option would probably be to make an Indian launch pad at the Arianespace site in Kourou in French Guiana to launch PSLVs and GSLVs.

The winds are in the sails, there is little doubt. As it is, Vikram Sarabhai's early work received so much political patronage because

A helicopter crash that captivated India. The Chief Minister of Andhra Pradesh YSR Reddy lost his life when a helicopter ferrying him crashed in the forests in 2009. This image taken by an ISRO asset shows the ill-fated helicopter crash.

Credit: ISRO

Is the Moon within reach? Young Indian entrepreneurs dream of reaching the Moon. Team Indus, a young bunch of enthusiastic space buffs preparing to launch a lunar satellite and rover atop India's Polar Satellite Launch Vehicle seen here in Bangalore.

Credit: Pallava Bagla

Rahul Narayan, one of the co-founders of Team Indus, calls himself a serial entrepreneur.

Credit: Pallava Bagla

of his intensive and never-say-die advocacy and efforts. This support has remained fairly steadfast over the decades, and has melted party lines. PM Modi has highlighted that 'continued progress in space must remain a national mission. We must keep enhancing our space capabilities. We must develop more advanced satellites; with higher computing, imaging and transmitting power. We must expand our satellite footprint, in terms of frequency and quality. We must also strengthen our international partnerships in all areas of space technology.' The PM's challenge to ISRO on developing a satellite jointly with SAARC nations is also being taken seriously. All in all, it is hardly surprising that,

Future bright. Indian children dressed as astronauts at a space conference in Hyderabad.

Credit: Pallava Bagla

to Radhakrishnan, the Mars mission fits into a much larger picture of the never-ceasing exploration of space, an exploration that must be and will always remain people-centric, with the vision to locate solutions to the problems of ordinary people and society. And in the infiniteness that is the Universe, people always find their peace. One of the Shanti Mantras of the Hindu Upanishads says:

ॐ पूर्णमदः पूर्णमिदम् पूर्णात् पूर्णमुदच्यते।
पूर्णस्य पूर्णमादाय पूर्णमेवावशिष्यते।।
ॐ शान्तिः शान्तिः शान्तिः

Om! That is infinite, and this (Universe) is infinite.
The infinite proceeds from the infinite.
(Then) taking the infinitude of the infinite (Universe),
It remains as the infinite alone.
Om! Peace! Peace! Peace!

Overall, the first inter-planetany mission has brought a new challenge as part of this infinite journey of space exploration. It is the search for a challenge and its solutions that human endeavour lives for and thrives on. This call of the frontier is what keeps us going, regardless of success or failure. Mahatma Gandhi had said: 'Glory lies in the attempt to reach one's goal and not in reaching it.'

NOTES

1. Press Information Bureau. Government of India, Department of Space: 9.7.14.
2. 12th Five Year Plan: Report of the Working Group of the Department of Space, WG 14.
3. Bagla, P. February 15, 2014. *ISRO unveils space capsule that will fly Indian astronauts.* http://www.ndtv.com/article/india/isro-unveils-space-capsule-that-will-fly-indian-astronauts-483605. Accessed on July 20, 2014.
4. Press Trust of India. January 8, 2014.
5. http://www.isro.gov.in/scripts/futureprogramme.aspx; accessed on July 12, 2014.
6. Bagla, P. and Menon, S. 2008. Destination Moon: India's Quest for the Moon, Mars and Beyond. Harper Collins Publishers India.
7. http://www.isro.org/pressrelease/scripts/pressreleasein.aspx?Aug30_2010. Accessed on July 21, 2014.
8. http://isrohq.vssc.gov.in/isr0dem0v2/index.php/science/science-history/74-general/1010-chandrayaan-2; accessed on July 16, 2014.
9. K Radhakrishnan, Address at the Convocation of IIT-Kanpur, June 18, 2014.

Chand pe bhi daag hai, which means even the Moon has blemishes, in Hindi. Paeans have been written comparing the Moon to the beauty of the virgin. This image from NASA would make many women wince on being compared to Earth's nearest planetary neighbour.

Credit: NASA

Two cultures of India! Worshipping the Agni-5 missile, on Wheeler Island off the coast of Odisha! Here a Hindu priest performs a 'puja' or religious ceremony to seek divine intervention before the launch of this complex system in 2013. It must be said here that the engineers who have crafted the Indian missile also follow a quirky tradition: they break coconuts in front of the missile and if these break into two equal halves it is believed that due justice will be done for this 'weapon of peace', which is how Indian defence scientists prefer to call this potent weapon! A missile is essentially a rocket that can accomplish re-entry into the atmosphere. The Agni-5 is a three-in-one missile as it can launch nuclear warheads, shoot down enemy satellites and, if push comes to shove, also help launch small satellites. With this over 5,000 kilometre range missile, India joined a select club of nations like the United States of America, United Kingdom, Russia, France and China that have the capability to operate a missile across continents.

Credit: Pallava Bagla

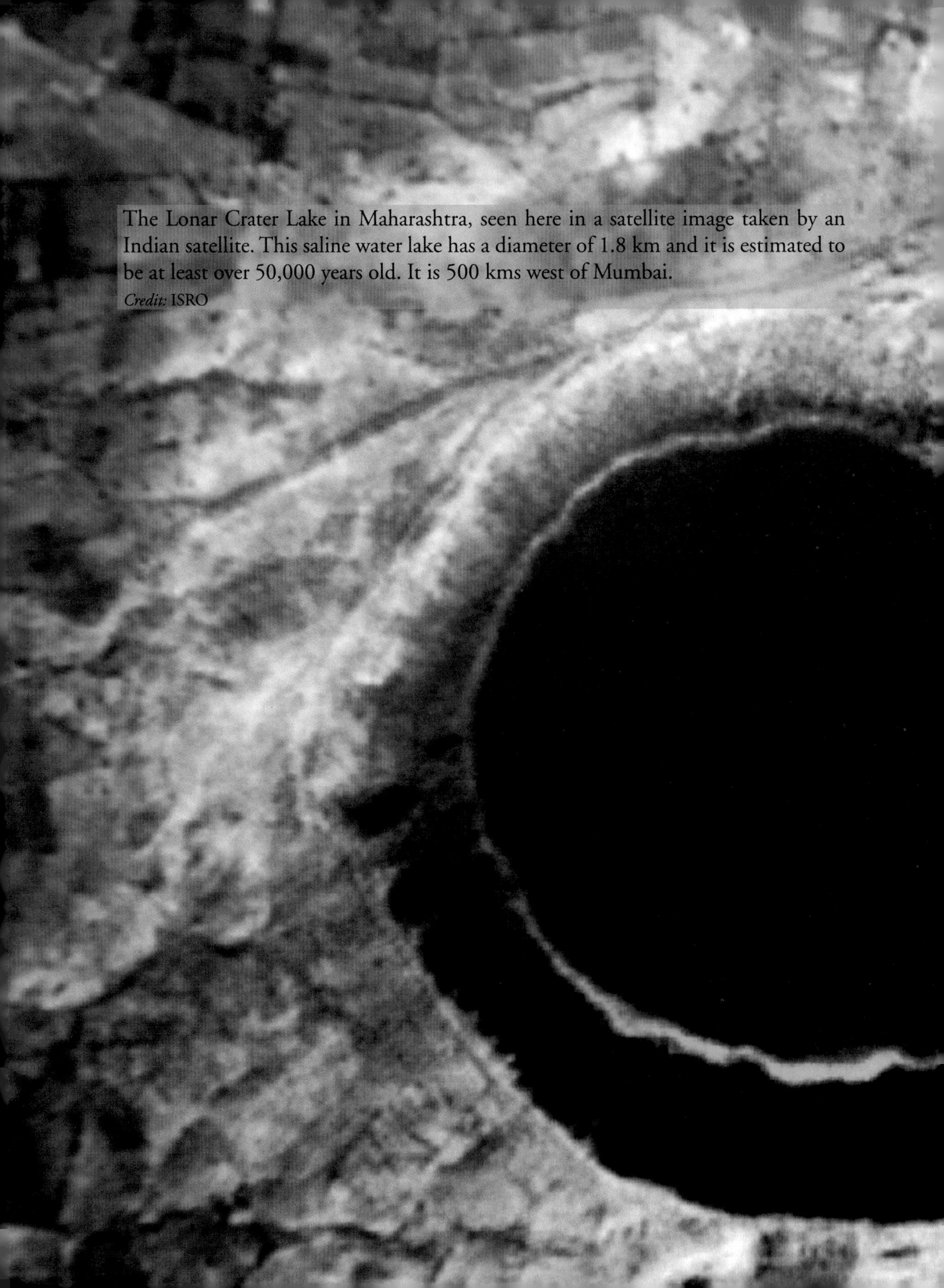

The Lonar Crater Lake in Maharashtra, seen here in a satellite image taken by an Indian satellite. This saline water lake has a diameter of 1.8 km and it is estimated to be at least over 50,000 years old. It is 500 kms west of Mumbai.
Credit: ISRO

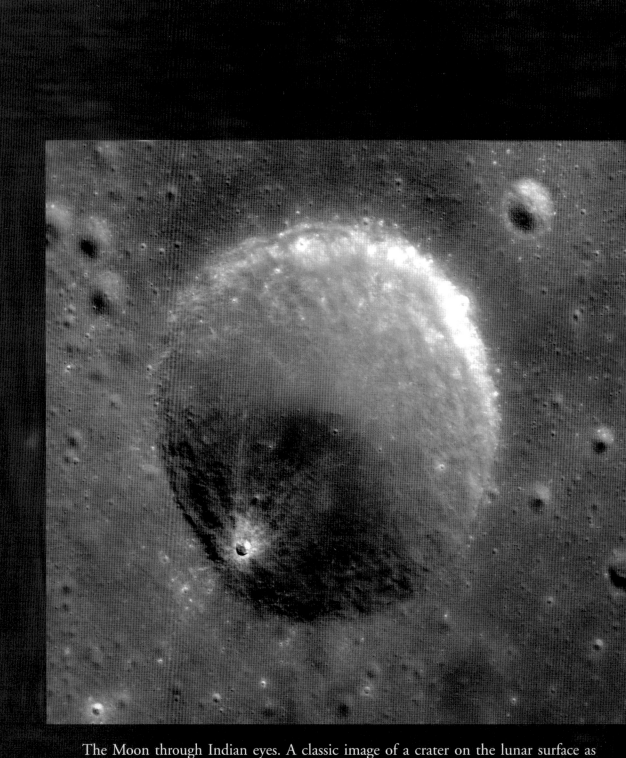

The Moon through Indian eyes. A classic image of a crater on the lunar surface as captured by Chandrayaan-1.
Credit: ISRO

Annexures

Milestones of the Indian Space Programme

Indian
Space Odyssey
1962-2014

Courtesy: ISRO

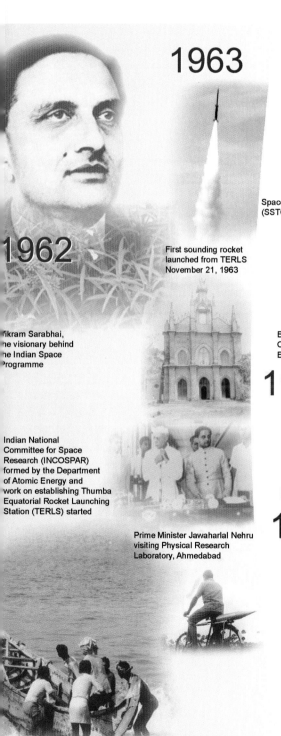

1963

1962

Vikram Sarabhai,
the visionary behind
the Indian Space
Programme

Indian National
Committee for Space
Research (INCOSPAR)
formed by the Department
of Atomic Energy and
work on establishing Thumba
Equatorial Rocket Launching
Station (TERLS) started

First sounding rocket
launched from TERLS
November 21, 1963

Prime Minister Jawaharlal Nehru
visiting Physical Research
Laboratory, Ahmedabad

1965

Space Science & Technology Centre
(SSTC) established in Thumba

1967

Experimental Satellite
Communication
Earth Station set up at Ahmedabad

1968

Prime Minister Indira Gandhi, dedicating
TERLS to U.N. February 2, 1968

1969

Formation of Indian Space Research
Organisation (ISRO) August 15, 1969

New SSTC campus at Veli,
Thiruvananthapuram

1971

SHAR Centre, Sriharikota operationalised
October 1971, renamed as Satish Dhawan
Space Centre in October 2003

1972

Department of Space (DOS)
established, ISRO brought under
DOS June 1, 1972

Prof. Satish Dhawan takes over as
Secretary, DOS & Chairman, ISRO
October 1972

Space Applications Centre (SAC)
established at Ahmedabad

ISRO Satellite Centre (ISAC)
established at Bangalore

1975

ISRO becomes Government Organisation (April 1, 1975)

Satellite Instructional Television Experiment (SITE) (1975-1976) using ATS-6 Satellite of USA

First Indian Satellite, ARYABHATA, launched on April 19, 1975

1977

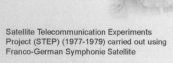

Satellite Telecommunication Experiments Project (STEP) (1977-1979) carried out using Franco-German Symphonie Satellite

1979

Launch of BHASKARA-I, an experimental satellite for earth observations (June 7, 1979)

First experimental launch of SLV-3 (August 10, 1979) The satellite did not reach orbit

1980

Liquid Propulsion Systems Centre (LPSC) started in Bangalore

Second experimental launch of SLV-3, Rohini RS-1 satellite placed in orbit (July 18, 1980)

1981

First developmental launch of SLV-3, RS-D1 placed in orbit (May 31, 1981)

Launch of APPLE, an experimental geo-stationary communication satellite (June 19, 1981)

Launch of BHASKARA-II (November 20, 1981)

1982

Master Control Facility (MCF) established at Hassan

Launch of INSAT-1A (April 10, 1982)

1983

Imagery from RS-D2

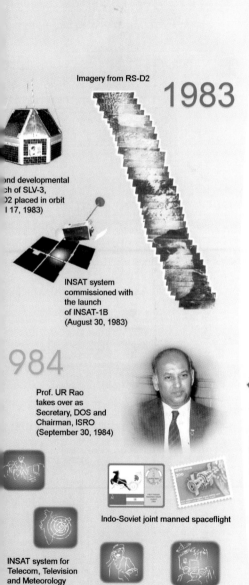

...ond developmental
...ch of SLV-3,
...02 placed in orbit
...17, 1983)

INSAT system
commissioned with
the launch
of INSAT-1B
(August 30, 1983)

984

Prof. UR Rao
takes over as
Secretary, DOS and
Chairman, ISRO
(September 30, 1984)

Indo-Soviet joint manned spaceflight

INSAT system for
Telecom, Television
and Meteorology

1987

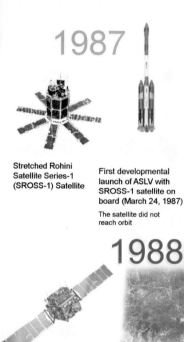

Stretched Rohini
Satellite Series-1
(SROSS-1) Satellite

First developmental
launch of ASLV with
SROSS-1 satellite on
board (March 24, 1987)

The satellite did not
reach orbit

1988

Imagery from IRS-1A

Establishment of IRS system
with the launch of Indian
Remote Sensing Satellite,
IRS-1A (March 17, 1988)

Second developmental
launch of ASLV with
SROSS-2 on board
(July 13, 1988)
The Satellite did not reach orbit

Launch of INSAT-1C
(July 22, 1988)

1990

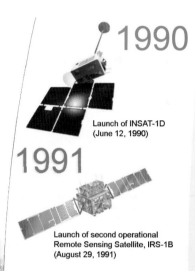

Launch of INSAT-1D
(June 12, 1990)

1991

Launch of second operational
Remote Sensing Satellite, IRS-1B
(August 29, 1991)

First successful
launch of ASLV,
SROSS-C placed in
orbit (May 20, 1992)

1992

SROSS-C Satellite

Launch of INSAT-2A, the first
satellite of the indigenously
built second generation INSAT
series (July 10, 1992)

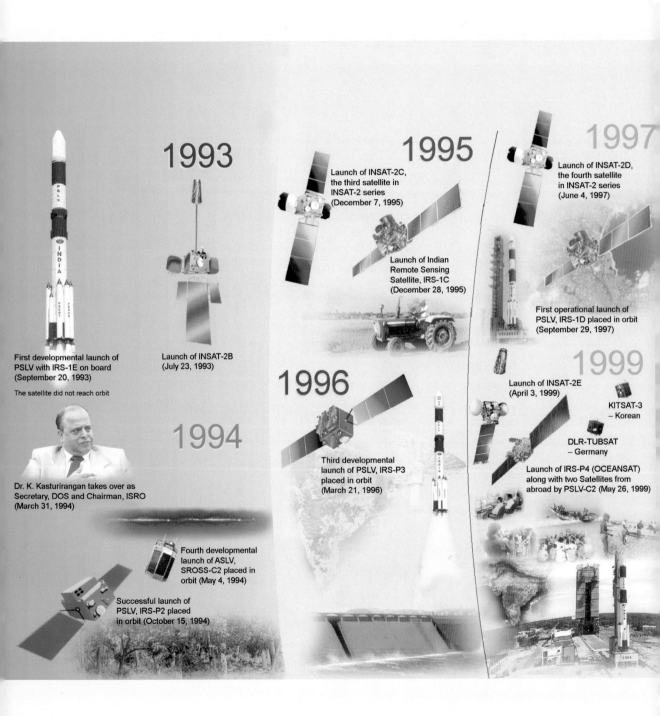

1993

First developmental launch of
PSLV with IRS-1E on board
(September 20, 1993)

The satellite did not reach orbit

Launch of INSAT-2B
(July 23, 1993)

Dr. K. Kasturirangan takes over as
Secretary, DOS and Chairman, ISRO
(March 31, 1994)

1994

Fourth developmental
launch of ASLV,
SROSS-C2 placed in
orbit (May 4, 1994)

Successful launch of
PSLV, IRS-P2 placed
in orbit (October 15, 1994)

1995

Launch of INSAT-2C,
the third satellite in
INSAT-2 series
(December 7, 1995)

Launch of Indian
Remote Sensing
Satellite, IRS-1C
(December 28, 1995)

1996

Third developmental
launch of PSLV, IRS-P3
placed in orbit
(March 21, 1996)

1997

Launch of INSAT-2D,
the fourth satellite
in INSAT-2 series
(June 4, 1997)

First operational launch of
PSLV, IRS-1D placed in orbit
(September 29, 1997)

1999

Launch of INSAT-2E
(April 3, 1999)

KITSAT-3
– Korean

DLR-TUBSAT
– Germany

Launch of IRS-P4 (OCEANSAT)
along with two Satellites from
abroad by PSLV-C2 (May 26, 1999)

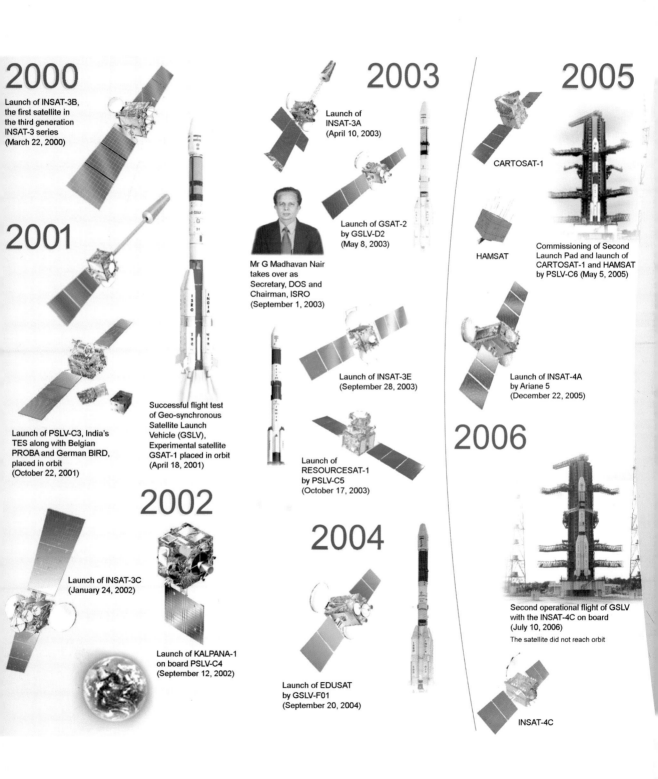

2000

Launch of INSAT-3B, the first satellite in the third generation INSAT-3 series (March 22, 2000)

2001

Launch of PSLV-C3, India's TES along with Belgian PROBA and German BIRD, placed in orbit (October 22, 2001)

Successful flight test of Geo-synchronous Satellite Launch Vehicle (GSLV), Experimental satellite GSAT-1 placed in orbit (April 18, 2001)

2002

Launch of INSAT-3C (January 24, 2002)

Launch of KALPANA-1 on board PSLV-C4 (September 12, 2002)

2003

Launch of INSAT-3A (April 10, 2003)

Launch of GSAT-2 by GSLV-D2 (May 8, 2003)

Mr G Madhavan Nair takes over as Secretary, DOS and Chairman, ISRO (September 1, 2003)

Launch of INSAT-3E (September 28, 2003)

Launch of RESOURCESAT-1 by PSLV-C5 (October 17, 2003)

2004

Launch of EDUSAT by GSLV-F01 (September 20, 2004)

2005

CARTOSAT-1

HAMSAT

Commissioning of Second Launch Pad and launch of CARTOSAT-1 and HAMSAT by PSLV-C6 (May 5, 2005)

Launch of INSAT-4A by Ariane 5 (December 22, 2005)

2006

Second operational flight of GSLV with the INSAT-4C on board (July 10, 2006)

The satellite did not reach orbit

INSAT-4C

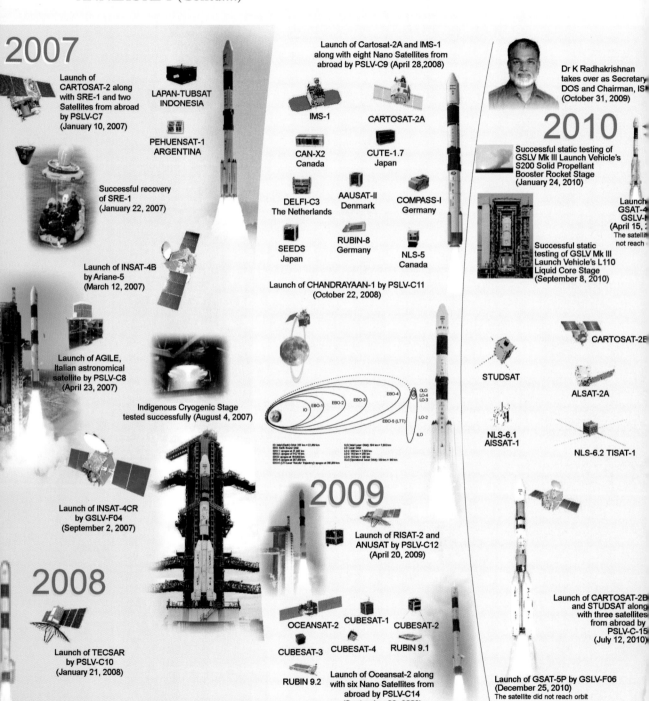

2007

Launch of CARTOSAT-2 along with SRE-1 and two Satellites from abroad by PSLV-C7 (January 10, 2007)

LAPAN-TUBSAT INDONESIA

PEHUENSAT-1 ARGENTINA

Successful recovery of SRE-1 (January 22, 2007)

Launch of INSAT-4B by Ariane-5 (March 12, 2007)

Launch of AGILE, Italian astronomical satellite by PSLV-C8 (April 23, 2007)

Indigenous Cryogenic Stage tested successfully (August 4, 2007)

Launch of INSAT-4CR by GSLV-F04 (September 2, 2007)

2008

Launch of TECSAR by PSLV-C10 (January 21, 2008)

Launch of Cartosat-2A and IMS-1 along with eight Nano Satellites from abroad by PSLV-C9 (April 28, 2008)

IMS-1

CARTOSAT-2A

CAN-X2 Canada

CUTE-1.7 Japan

DELFI-C3 The Netherlands

AAUSAT-II Denmark

COMPASS-I Germany

SEEDS Japan

RUBIN-8 Germany

NLS-5 Canada

Launch of CHANDRAYAAN-1 by PSLV-C11 (October 22, 2008)

2009

Launch of RISAT-2 and ANUSAT by PSLV-C12 (April 20, 2009)

OCEANSAT-2

CUBESAT-1

CUBESAT-2

CUBESAT-3

CUBESAT-4

RUBIN 9.1

RUBIN 9.2

Launch of Oceansat-2 along with six Nano Satellites from abroad by PSLV-C14 (September 23, 2009)

Dr K Radhakrishnan takes over as Secretary, DOS and Chairman, ISRO (October 31, 2009)

2010

Successful static testing of GSLV Mk III Launch Vehicle's S200 Solid Propellant Booster Rocket Stage (January 24, 2010)

Launch of GSAT-4 by GSLV-D3 (April 15, 2010) The satellite did not reach orbit

Successful static testing of GSLV Mk III Launch Vehicle's L110 Liquid Core Stage (September 8, 2010)

CARTOSAT-2B

STUDSAT

ALSAT-2A

NLS-6.1 AISSAT-1

NLS-6.2 TISAT-1

Launch of CARTOSAT-2B and STUDSAT along with three satellites from abroad by PSLV-C-15 (July 12, 2010)

Launch of GSAT-5P by GSLV-F06 (December 25, 2010) The satellite did not reach orbit

2011

X-SAT   YOUTHSAT   RESOURCESAT-2

Launch of RESOURCESAT-2, YOUTHSAT and X-SAT
by PSLV-C16 (April 20, 2011)

Launch of GSAT-8
by Ariane 5 VA-202
(May 21, 2011)

GSAT-12

Launch of GSAT-12 by
PSLV-C17 (July 15, 2011)

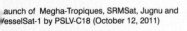

MEGHA-TROPIQUES

SRMSat   VesselSat-1   JUGNU

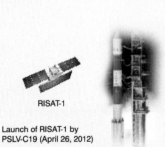

Launch of Megha-Tropiques, SRMSat, Jugnu and
VesselSat-1 by PSLV-C18 (October 12, 2011)

2012

RISAT-1

Launch of RISAT-1 by
PSLV-C19 (April 26, 2012)

PROITERES   SPOT 6

Launch of SPOT 6 and PROITERES
by PSLV-C21 (September 09, 2012)

Launch of GSAT-10
by Ariane 5 VA-209
(September 29, 2012)

2013

SAPPHIRE   NEOSSAT

SARAL

NLS 8.1

NLS 8.2   NLS 8.3   STRaND-1

Launch of SARAL, SAPPHIRE, NEOSSAT, NLS 8.1,
NLS 8.2, NLS 8.3 and STRaND-1 by PSLV-C20
(February 25, 2013)

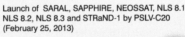

Launch of IRNSS-1A
by PSLV-C22
(July 1, 2013)

Launch of INSAT-3D
by Ariane 5 VA-214
(July 26, 2013)

Launch of GSAT-7
by Ariane 5 VA-215
(August 30, 2013)

Launch of
MARS ORBITER MISSION
by PSLV-C25 (November 5, 2013)

2014

GSAT-14

Launch of GSAT-14 by GSLV-D5
(January 05, 2014)

IRNSS-1B

Launch of IRNSS-1B by PSLV-C24
(APRIL 04, 2014)

FUTURE...

ASTROSAT

CARTOSAT-2C/2D

RLV

INDIAN REGIONAL
NAVIGATION
SATELLITE SYSTEM
(IRNSS)

GSAT-11

CHANDRAYAAN-2

ADITYA

HUMAN SPACE FLIGHT

GSLV-MkIII

The Mangalyaan, deconstructed

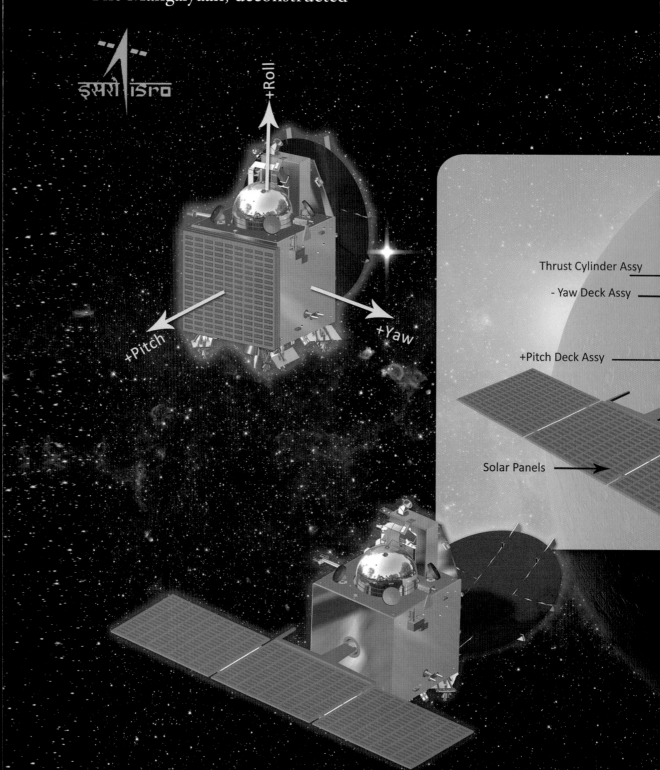

+Roll

+Pitch

+Yaw

Thrust Cylinder Assy

- Yaw Deck Assy

+Pitch Deck Assy

Solar Panels

इसरो ISRO

Various views of the
Mars Orbiter Mission

Credit: ISRO

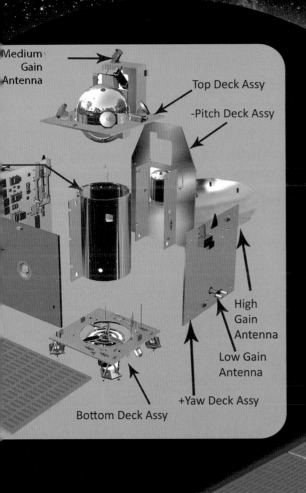

Medium Gain Antenna

Top Deck Assy

-Pitch Deck Assy

High Gain Antenna

Low Gain Antenna

+Yaw Deck Assy

Bottom Deck Assy

Orbital Callisthenics: Challenges of an Inter-planetary Mission

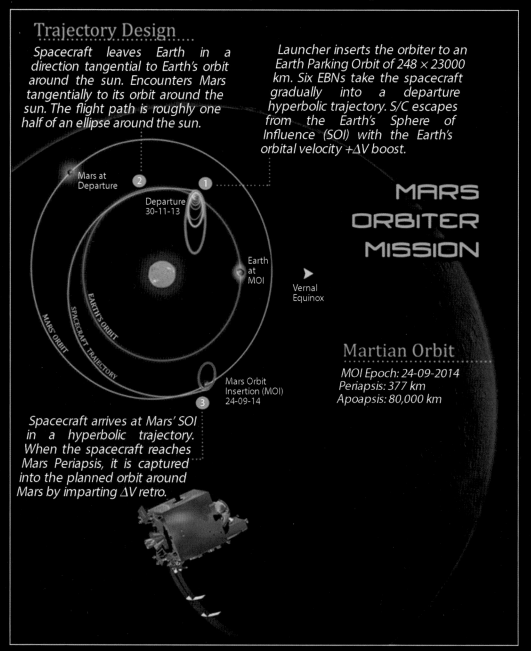

Trajectory Design

Spacecraft leaves Earth in a direction tangential to Earth's orbit around the sun. Encounters Mars tangentially to its orbit around the sun. The flight path is roughly one half of an ellipse around the sun.

Launcher inserts the orbiter to an Earth Parking Orbit of 248 × 23000 km. Six EBNs take the spacecraft gradually into a departure hyperbolic trajectory. S/C escapes from the Earth's Sphere of Influence (SOI) with the Earth's orbital velocity +ΔV boost.

Mars at Departure

2   1

Departure
30-11-13

Earth
at
MOI

▶
Vernal
Equinox

MARS
ORBITER
MISSION

EARTH'S ORBIT

SPACECRAFT TRAJECTORY

MARS ORBIT

SPACECRAFT TRAJECTORY

Mars Orbit
Insertion (MOI)
24-09-14

3

Martian Orbit

*MOI Epoch: 24-09-2014
Periapsis: 377 km
Apoapsis: 80,000 km*

Spacecraft arrives at Mars' SOI in a hyperbolic trajectory. When the spacecraft reaches Mars Periapsis, it is captured into the planned orbit around Mars by imparting ΔV retro.

The road map for Mars. The trajectory or path of India's Mars Orbiter Mission after its launch from Sriharikota through its nine month journey

Orbital Callisthenics: Challenges of an Inter-planetary Mission

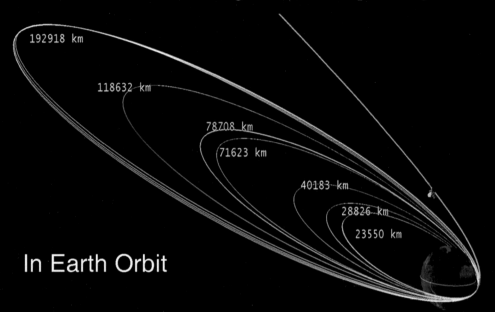

In Earth Orbit

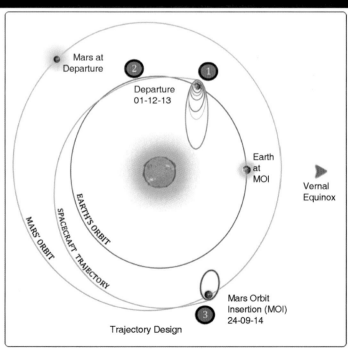

Trajectory of the Mars Orbiter Mission.

Asking the right questions: MOM's Payloads

MARS
ORBITER
MISSION

Payloads

Lyman Alpha Photometer (LAP)

Lyman Alpha Photometer (LAP) is an absorption cell photometer. It measures the relative abundance of deuterium and hydrogen from Lyman-alpha emission in the Martian upper atmosphere (typically Exosphere and exobase). Measurement of D/H (Deuterium to Hydrogen abundance Ratio) allows us to understand especially the loss process of water from the planet.

Methane Sensor for Mars (MSM)

MSM is designed to measure Methane (CH_4) in the Martian atmosphere with PPB accuracy and map its sources. Data is acquired only over illuminated scene as the sensor measures reflected solar radiation. Methane concentration in the Martian atmosphere undergoes spatial and temporal variations.

Mars Exospheric Neutral Composition Analyser (MENCA)

MENCA is a quadruple mass spectrometer capable of analysing the neutral composition in the range of 1 to

Atmospheric studies

Plasma and partic

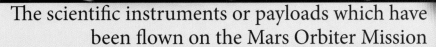

The scientific instruments or payloads which have been flown on the Mars Orbiter Mission

Courtesy: ISRO

Mars Color Camera (MCC)

This tri-color Mars Color camera gives images and information about the surface features and composition of Martian surface. They are useful to monitor the dynamic events and weather of Mars. MCC will also be used for probing the two satellites of Mars – Phobos and Deimos. It also provides the context information for other science payloads.

Thermal Infrared Imaging Spectrometer (TIS)

TIS measures the thermal emission and can be operated during both day and night. Temperature and emissivity are the two basic physical parameters estimated from thermal emission measurement. Many minerals and soil types have characteristic spectra in TIR region. TIS can map surface composition and mineralogy of Mars.

300 amu with unit mass resolution. The heritage of this payload is from Chandra's Altitudinal Composition Explorer (CHACE) payload.

vironment studies

Surface Imaging Studies

The Polar Satellite Launch Vehicle: Inside Story

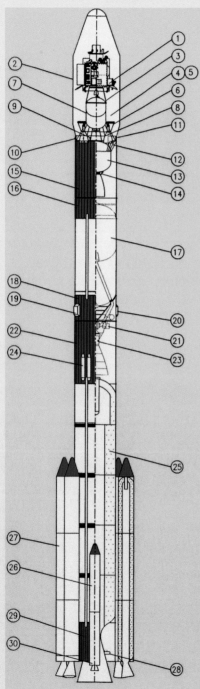

1. MANGALYAAN
2. HEATSHIELD
3. PAYLOAD ADAPTOR
4. EQUIPMENT BAY
5. FOURTH STAGE PROPELLANT TANK
6. FOURTH STAGE ENGINE
7. ANTENNAE
8. REACTION CONTOL THRUSTER
9. INTERSTAGE
10. THIRD STAGE ADAPTOR
11. THIRD STAGE MOTOR
12. FLEX NOZZLE CONTROL SYSTEM
13. INTERSTAGE
14. INTERSTAGE
15. SECOND STAGE PROPELLANT TANK
16. INTERSTAGE
17. SECOND STAGE RETRO ROCKET
18. ULLAGE ROCKET
19. GIMBAL CONTROL SYSTEM
20. INTERSTAGE
21. SECOND STAGE ENGINE
22. FIRST STAGE RETRO ROCKET
23. FIRST STAGE MOTOR
24. SITVC INJECTANT TANK
25. STRAP-ON MOTOR (XL)
26. SITVC SYSTEM
27. CORE BASE SHROUND
28. ROLL CONTROL ENGINE

OVERALL HEIGHT – 44.4 METRES
LIFT OFF WEIGHT – 320 TONS

Complexities of the Polar Satellite Launch Vehicle

The World's Love Affair with Mars

A Global Chronology of Mars Exploration

Year	Country	Mission Timeline
1960		Marsnik 1 (Mars 1960A) – 10 October 1960 – Attempted Mars Flyby (Launch Failure)
		Marsnik 2 (Mars 1960B) – 14 October 1960 – Attempted Mars Flyby (Launch Failure)
1962		Sputnik 22 – 24 October 1962 – Attempted Mars Flyby
		Mars 1 – 1 November 1962 – Mars Flyby (Contact Lost)
		Sputnik 24 – 4 November 1962 – Attempted Mars Lander
1964		Mariner 3 – 5 November 1964 – Attempted Mars Flyby
		Mariner 4 – 28 November 1964 – Mars Flyby
		Zond 2 – 30 November 1964 – Mars Flyby (Contact Lost)
1965		Zond 3 – 18 July 1965 – Lunar Flyby, Mars Test Vehicle
1969		Mariner 6 – 25 February 1969 – Mars Flyby
		Mariner 7 – 27 March 1969 – Mars Flyby
		Mars 1969A – 27 March 1969 – Attempted Mars Orbiter (Launch Failure)
		Mars 1969B – 2 April 1969 – Attempted Mars Orbiter (Launch Failure)
1971		Mariner 8 – 8 May 1971 – Attempted Mars Flyby (Launch Failure)
		Cosmos 419 – 10 May 1971 – Attempted Mars Orbiter/Lander
		Mars 2 – 19 May 1971 – Mars Orbiter/Attempted Lander
		Mars 3 – 28 May 1971 – Mars Orbiter/Lander
		Mariner 9 – 30 May 1971 – Mars Orbiter
1973		Mars 4 – 21 July 1973 – Mars Flyby (Attempted Mars Orbiter)
		Mars 5 – 25 July 1973 – Mars Orbiter
		Mars 6 – 5 August 1973 – Mars Lander (Contact Lost)
		Mars 7 – 9 August 1973 – Mars Flyby (Attempted Mars Lander)
1975		Viking 1 – 20 August 1975 – Mars Orbiter and Lander
		Viking 2 – 9 September 1975 – Mars Orbiter and Lander
1988		Phobos 1 – 7 July 1988 – Attempted Mars Orbiter/Phobos Landers
		Phobos 2 – 12 July 1988 – Mars Orbiter/Attempted Phobos Landers
1992		Mars Observer – 25 September 1992 – Attempted Mars Orbiter (Contact Lost)

Year	Country	Mission Timeline
1996		Mars Global Surveyor – 7 November 1996 – Mars Orbiter
		Mars 96 – 16 November 1996 – Attempted Mars Orbiter/Landers
		Mars Pathfinder – 4 December 1996 – Mars Lander and Rover
1998		Nozomi (Planet-B) – 3 July 1998 – Mars Orbiter
		Mars Climate Orbiter – 11 December 1998 – Attempted Mars Orbiter
1999		Mars Polar Lander – 3 January 1999 – Attempted Mars Lander
		Deep Space 2 (DS2) – 3 January 1999 – Attempted Mars Penetrators
2001		2001 Mars Odyssey – 7 April 2001 – Mars Orbiter
2003		Mars Express – 2 June 2003 – Mars Orbiter and Lander
		Spirit (MER-A) – 10 June 2003 – Mars Rover
		Opportunity (MER-B) – 7 July 2003 – Mars Rover
2005		Mars Reconnaissance Orbiter – 10 August 2005 – Mars Orbiter
2007		Phoenix – 4 August 2007 – Mars Scout Lander
2011		Phobos-Grunt – 8 November 2011 – Attempted Martian Moon Phobos Lander
		Yinghuo-1 – 8 November 2011 – Attempted Mars Orbiter
		Mars Science Laboratory – 26 November 2011 – Mars Rover
2013		Mars Orbiter Mission or Mangalyaan – 5 November 2013 – Mars Orbiter
		MAVEN – 18 November 2013 – Mars Scout Mission Orbiter

Source: NASA

LESSONS LEARNT FROM OTHER MARS MISSIONS

Mission Type	Success Rate	Total Attempts	Success	Partial Success	Launch Failure	Failed Enroute	Failed to Orbit/Land
Flyby	45%	11	5		4	2	
Orbiter	50%	22	9	2	5	3	3
Lander	30%	10	3			3	4
Rover	57%	7	4	1			2
Sample Return	0%	1				1	
Total	42%	51	21	3	9	9	9

Majority of failures are primarily due to launch related issues followed by propulsion system problems, software errors both in ground and on-board, human error, insufficient hardware testing and the conceived mission concepts.

Source: ISRO

For the day when Indians will travel to space

As of 2007

Indian Human Space Program

Courtesy: ISRO

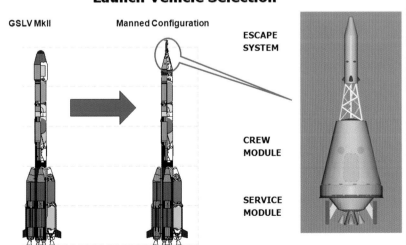

Launch Vehicle Selection

GSLV MkII   Manned Configuration

ESCAPE SYSTEM

CREW MODULE

SERVICE MODULE

> P/L in LEO : 5500kg; P/L after man rating: 4450 kg
> Launch Azimuth: 140° from SLP, SDSC/SHAR
> P/L requirement: 4000 kg; Adequate margin for mass growth
> GSLV MKII is operational and needs minimal changes

Schematics of the Indian Human Space Programme

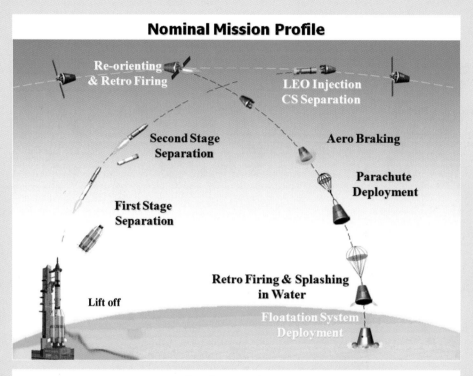

Nominal Mission Profile

Re-orienting & Retro Firing

LEO Injection CS Separation

Second Stage Separation

Aero Braking

First Stage Separation

Parachute Deployment

Retro Firing & Splashing in Water

Lift off

Floatation System Deployment

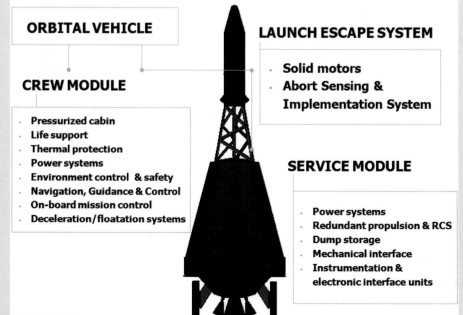

Manned Mission – New Systems

Human Space Flight: Critical Technology Development

ORBITAL VEHICLE

LAUNCH ESCAPE SYSTEM

- Solid motors
- Abort Sensing & Implementation System

CREW MODULE

- Pressurized cabin
- Life support
- Thermal protection
- Power systems
- Environment control & safety
- Navigation, Guidance & Control
- On-board mission control
- Deceleration/floatation systems

SERVICE MODULE

- Power systems
- Redundant propulsion & RCS
- Dump storage
- Mechanical interface
- Instrumentation & electronic interface units

Crew module – Salient features

Mass : 3000 kg

❑ Provision for 2 crew; extendable to 3 crew with

additional mass of 200 kg

Houses crew cabin and equipments

❑ Twin-walled construction

❑ Controlled guided flight

❑ Lift to drag ratio: 0.20

❑ Max. stagnation heat flux: < 70 W/cm^2

❑ Maximum deceleration: less than 4 g

Launch Escape System [LES] & Test Vehicle

Launch Escape System

Launch Escape System

To pull crew module clear of launch vehicle

- Automatic Fault detection and activation
- Solid motor for quick action
- Separated and jettisoned in nominal mission
- High reliability
- Mass : 1000 kg

Furthering Astronomical Research

ASTROSAT

A Multi-wavelength Astronomical Space Observatory

ASTROSAT is a multi-wavelength space observatory dedicated for astronomical research. The satellite will have simultaneous wide spectral coverage extending over Visible, Ultraviolet, soft X-ray and hard X-ray and low energy Gamma ray regions. The 1500 kg ASTROSAT is scheduled to be launched by the India's workhorse launch vehicle PSLV into a 650 km high orbit with 8-degree orbital inclination.

Major Scientific Objectives of ASTROSAT are :

- Observations of Celestial Bodies and Cosmic Sources in X-Ray and UV Spectral bands
- Monitoring X-ray sky for new transients
- All sky survey in the hard X-ray and UV bands
- Broad band spectroscopic studies of clusters of galaxies and stellar coronae

The payloads carried by ASTROSAT are:

- Three Large Area Xenon-filled Proportional Counters in 3-100 keV band for timing and spectral studies
- Cadmium Zinc Telluride array with coded mask aperture for hard X-ray imaging and spectral studies in 10-100 keV
- Soft X-ray Telescope with CCD camera for timing and variability studies in the X-ray bandwidth of 0.3 to 10 keV
- Scanning X-ray Sky Monitor for timescales and luminosity variations in 2-10 keV using proportional counter system
- Ultra Violet Imaging telescope to cover visible, near ultra violet and far ultra violet bands in130-600 nm bands
- Charge Particle Monitor with a 10 mm cube of Cesium Iodide (Tellurium) CsI (TI) crystal viewed by a Photodiode for detecting high-energy particles in the satellite orbital path and for alerting the instrumentation from possible damage

Useful Further Reading

PRINTED AND ONLINE SOURCES

1. Lele, A. Mission Mars: India's Quest for the Red Planet. Springer Briefs in Applied Science and Technology, Springer, 2014.

2. Rao, UR. India's Rise as a Space Power. Foundation Books, 2014.

3. Laxman, S. Mars Beckons India: The Story of India's Mission to Mars. Vigyan Prasar, 2012.

4. Pyle, R. Destination Mars: New Explorations of the Red Planet. Prometheus Books, 2012.

5. Das, SK. Destination Mars: Secrets of the Red Planet Revealed. Rupa Publications India (Red Turtle), 2013.

6. Bagla, P & Menon, S. Destination Moon: India's Quest for the Moon, Mars and Beyond. Harper Collins Publishers, 2008.

7. Bagla, P & Menon, S. Mission Moon (Hindi). Prabhat Prakashan, New Delhi, 2009.

8. Manoranjan Rao, PV & Radhakrishnan P. A Brief History of Rocketry in ISRO. Universities Press, 2012.

9. Jain, G, Sharma, S, Ajai. Captivating Views of India. Space Applications Centre, ISRO, 2013.

10. Sarabhai, VA. Sources of Man's Knowledge. National Programme of Talks Series, Bangalore, Resonance, December 2001.

11. Bagla, P. A tiny step for a giant leap? The Hindu, October 31, 2013 (http://www.thehindu.com/opinion/op-ed/a-tiny-step-for-a-giant-leap/article5297789.ece).

12. Hogan, T. Mars Wars: The Rise and Fall of the Space Exploration Initiative. The NASA History Series, NASA, 2007.

13. Chandrayaan-1: Lunar Science Atlas, Space Applications Centre, ISRO, 2013.

14. Bagla, P. India Sets its Sights on Mars. Physics World, December, 2012.

15. Space Science in the 21st Century: A Task Group Report. National Academies Press, Washington, DC, 1998.

16. Connor, S. Desperately Seeking Life on Mars. Space Policy, 2002: 18: 267–269.

17. Sadeh, E. (Ed). Politics of Space. Routledge, London, 2011, p. 3.

18. Mapping Mars: Science, Imagination, and the Birth of a World. Picador USA: New York, p. 98.

19. Sankar, U. The Economics of India's Space Programme. Oxford University Press, New Delhi, 2007, 1–2.

20. Manoranjan Rao, PV. No Ambiguity of Purpose: The Indian Space Programme. In: Manoranjan Rao, PV (Ed), 50 Years of Space: A Global Perspective. Astronomical Society of India, Universities Press (India), 2007.

21. Lele, A. Asian Space Race: Rhetoric or Reality? Springer, Heidelberg, 2013, pp. 59–67. India's Space Programme.

22. Bharath, G. Space Security: India, in Crux of Asia. In: Tellis, A., Mirski, S., Editors. Carnegie Endowment for International Peace, 2013.

23. Lowell, Percival. Mars. Cheshire, England: New Line Publishing, 2009. First published in 1897, in New York, by Houghton Mifflin.

24. National Research Council. Pathways to Exploration – Rationales and Approaches for a US Program of Human Space Exploration. Washington DC, The National Academies Press, 2014.

25. www.isro.gov.in

26. https://www.facebook.com/isroofficial

27. https://www.facebook.com/isromom

28. https://twitter.com/ISROOFFICIAL

29. www.ndtv.com/mars

30. http://www.ndtv.com/topic/mission-moon

31. www.nasa.gov

32. www.esa.int

33. www.jpl.nasa.gov

34. A. Shah. Vikram Sarabhai: A Life. Penguin Viking. 2007.

35. Bagla, P. Coverage of India's Mission Mars on NDTV (2013-2014): Links on YouTube: (http://www.youtube.com/watch?v=UgLVegY9Bqs&list=PLYSfYVdrOZvgf_CfpSWMHkfz-iCyPU22Y):

    Indo-US space relation ready for lift-off: NASA chief General Charles Bolden to NDTV
    http://www.ndtv.com/video/player/ndtv-special-ndtv-24x7/indo-us-space-relation-ready-for-lift-off-nasa-chief-general-charles-bolden-to-ndtv/288873

    India's Mars Mission: First look at the satellite that will orbit the planet
    http://www.ndtv.com/video/player/news/india-s-mars-mission-first-look-at-the-satellite-that-will-orbit-the-planet/290208

    India's Mars mission: Mangalyaan to begin its 10-month journey on Oct 28
    http://www.ndtv.com/video/player/news/india-s-mars-mission-mangalyaan-to-begin-its-10-month-journey-on-oct-28/291550

India's Mars Mission is all about technology (Aired: Jan 4, 2013)

http://www.ndtv.com/video/player/news/india-s-mars-mission-is-all-about-technology-aired-jan-4-2013/293615

Mangalyaan is world's cheapest inter-planetary mission

http://www.ndtv.com/video/player/news/mangalyaan-is-world-s-cheapest-inter-planetary-mission/293735

'Mars will be colonized soon'

http://www.ndtv.com/video/player/news/mars-will-be-colonized-soon/293736

'Exciting that Mangalyaan may fly through a tail of comet'

http://www.ndtv.com/video/player/news/exciting-that-mangalyaan-may-fly-through-a-tail-of-comet/294282

Photographing Mars: Mangalyaan's special camera

http://www.ndtv.com/video/player/news/photographing-mars-mangalyaan-s-special-camera/294403

Mangalyaan is only the beginning, expect many more galactic leaps

http://www.ndtv.com/video/player/news/mangalyaan-is-only-the-beginning-expect-many-more-galactic-leaps/294526

Looking for signs of life on Mars

http://www.ndtv.com/video/player/news/looking-for-signs-of-life-on-mars/294630

Why is the Martian atmosphere so thin?

http://www.ndtv.com/video/player/news/why-is-the-martian-atmosphere-so-thin/294782

Destination Mars

http://www.ndtv.com/video/player/news/destination-mars/294807

Searching for hot and cold spots to map Martian resources

http://www.ndtv.com/video/player/news/searching-for-hot-and-cold-spots-to-map-martian-resources/294791

Mars is parched and hostile for life

http://www.ndtv.com/video/player/news/mars-is-parched-and-hostile-for-life/294792

With upcoming Mars mission, India looks to join elite club

http://www.ndtv.com/video/player/news/with-upcoming-mars-mission-india-looks-to-join-elite-club/295908

The making of 'Mangalyaan', India's satellite to Mars

http://www.ndtv.com/video/player/news/the-making-of-mangalyaan-india-s-satellite-to-mars/296146

First look at 'Mangalyaan', India's satellite to Mars

http://www.ndtv.com/video/player/news/first-look-at-mangalyaan-india-s-satellite-to-mars/296211

Mangalyaan a historic mission that will make India proud: ISRO chief to NDTV

http://www.ndtv.com/video/player/news/mangalyaan-a-historic-mission-that-will-make-india-proud-isro-chief-to-ndtv/296251

India is not racing with China; there is no Asian Space Race: ISRO chairman

http://www.ndtv.com/video/player/news/india-is-not-racing-with-china-there-is-no-asian-space-race-isro-chairman/296235

India's Mars mission: GSLV was the preferred rocket, says Indian scientist

http://www.ndtv.com/video/player/news/india-s-mars-mission-gslv-was-the-preferred-rocket-says-indian-scientist/296232

Is Mangalyaan all about an Asian space race?

http://www.ndtv.com/video/player/news/is-mangalyaan-all-about-an-asian-space-race/296384

Catching faint Martian whispers on Earth

http://www.ndtv.com/video/player/news/catching-faint-martian-whispers-on-earth/296490

Global excitement about India's Mars mission

http://www.ndtv.com/video/player/news/global-excitement-about-india-s-mars-mission/296491

Slept most nights in satellite centre: Mangalyaan project director

http://www.ndtv.com/video/player/news/slept-most-nights-in-satellite-centre-mangalyaan-project-director/296231

India's Mission Mars: The inside story

http://www.ndtv.com/video/player/ndtv-special-ndtv-24x7/india-s-mission-mars-the-inside-story/296582

This man found water on moon before finding it in his village

http://www.ndtv.com/video/player/news/this-man-found-water-on-moon-before-finding-it-in-his-village/296229

India already on Red Planet long before maiden Mars mission

http://www.ndtv.com/video/player/news/india-already-on-red-planet-long-before-maiden-mars-mission/296680

India's Mars mission: An inside view

http://www.ndtv.com/video/player/news/india-s-mars-mission-an-inside-view/296764

India's maiden Mars mission to be launched today

http://www.ndtv.com/video/player/news/india-s-maiden-mars-mission-to-be-launched-today/296780

India's maiden mission to Mars launched

http://www.ndtv.com/video/player/news/india-s-maiden-mission-to-mars-launched/296815

India's Mission Mars: science and sociology

http://www.ndtv.com/video/player/the-social-network/india-s-mission-mars-science-and-sociology/296838

Mars or malnutrition: India's priority?

http://www.ndtv.com/video/player/left-right-centre/mars-or-malnutrition-india-s-priority/296846

The ISRO dreamers behind India's Mars mission

http://www.ndtv.com/video/player/news/the-isro-dreamers-behind-india-s-mars-mission/296852

India mission to Mars successfully completes first stage

http://www.ndtv.com/video/player/news/india-mission-to-mars-successfully-completes-first-stage/296855

Conquering Mars

http://www.ndtv.com/video/player/the-buck-stops-here/conquering-mars/296856

Long journey to Mars, will take one step at a time: Dr Kiran Kumar

http://www.ndtv.com/video/player/news/long-journey-to-mars-will-take-one-step-at-a-time-dr-kiran-kumar/296861

ISRO chief on Mangalyaan's successful launch to NDTV

http://www.ndtv.com/video/player/news/isro-chief-on-mangalyaan-s-successful-launch-to-ndtv/296862

Mangalyaan 'pushed further away from earth' in midnight operation: Scientists

http://www.ndtv.com/video/player/news/mangalyaan-pushed-further-away-from-earth-in-midnight-operation-scientists/297072

PHOTOS

Mangalyaan, India's maiden satellite to Mars

http://www.ndtv.com/photos/news/mangalyaan-india-s-maiden-satellite-to-mars-15974

Space Odyssey 2013: India's Mission Mars

http://www.ndtv.com/photos/news/space-odyssey-2013-india-s-mission-mars-16109

First pics: Mangalyaan at space port

http://www.ndtv.com/photos/news/first-pics-mangalyaan-at-space-port-16132

Fourth rock from the sun

http://www.ndtv.com/photos/news/fourth-rock-from-the-sun-16137

The making of a rocket

http://www.ndtv.com/photos/news/the-making-of-a-rocket-16214

India's Mars Mission: Rocket blasts off

http://www.ndtv.com/photos/news/india-s-mars-mission-rocket-blasts-off-16350

The Moon in all its glory, depicted here in a composite image. Seen here are phases of the Moon's waxing and waning.

Credit: NASA

New Delhi

Kolkata

Mumbai

Hyderabad

Arabian Sea

Bay of Bengal

Bengaluru   Chennai

ISRO

INDIA AND ENVIRONS
viewed by Resourcesat-1